Transient Magnetic Fields

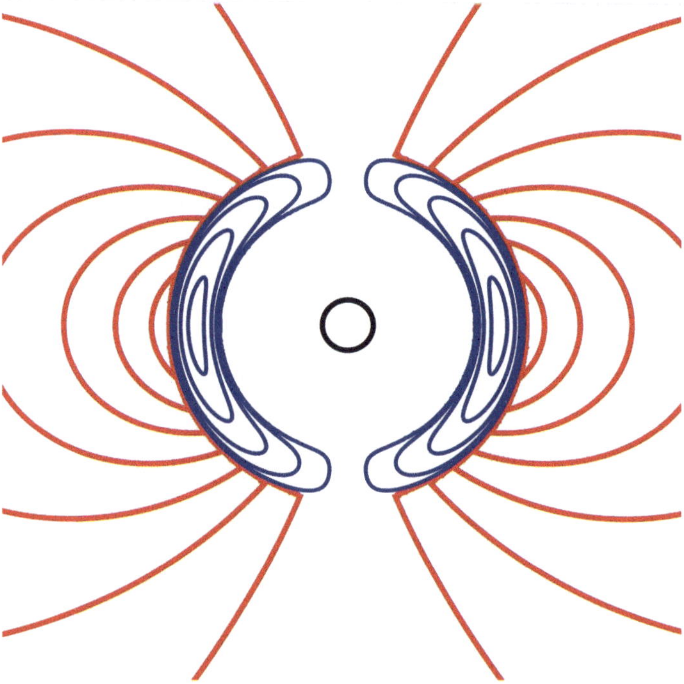

Fig. 0.1 Sky-plane cut through an axisymmetric dipole field (red), showing what happens to the field when its source current on the sphere (black circle) is turned off suddenly. The outgoing bubble of reversed polarity (blue) sweeps up the dipole field and carries its energy away as radiation, leaving no field behind

Neil R. Sheeley, Jr.

Transient Magnetic Fields

Springer

Neil R. Sheeley, Jr.
Alexandria
VA, USA

ISBN 978-3-030-40263-1 ISBN 978-3-030-40264-8 (eBook)
https://doi.org/10.1007/978-3-030-40264-8

This Springer imprint is published by the registered company Springer Nature Switzerland AG.
The registered company address is: Gewerbestrasse 11, 6330 Cham, Switzerland

To Marybeth

Preface

Some years ago, I thought that it would be interesting to see if I could find a model of how magnetic fields change with time when the displacement current is included in Maxwell's equations. I wanted to avoid infinities, like the field of an infinitely long, straight current, and the caveats that always accompany them. But I also wanted to keep the model simple, so that I could calculate the fields in closed form. I decided to try the current that produces the field of a uniformly magnetized sphere (a uniform field inside the sphere and a dipole field outside) and to ignore conducting material and the complications that accompany it. Thus, the model was like the potential field model that is often used in solar physics research, except that it included the electromagnetic waves and transients that are produced by the displacement current in Maxwell's equations. I thought that this might be a step toward understanding transient magnetic fields, even if Alfven waves and sound waves were excluded by omitting the conducting material.

I started with a simple step function for the current profile, and did the calculation at home, as a serious hobby. At first, I encountered what looked like a very nasty problem in geometry. However, as the reader will soon see, the nastiness unraveled and left a simple algebraic expression for the vector potential. This was a wonderful surprise because I had expected something much more complicated, perhaps involving spherical Bessel functions of one-half order (Hankel functions) like those that appear in many textbooks on electricity and magnetism.

The simple form of the vector potential made it easy to deduce the magnetic field (and the electric field of the wave) and many properties, including the radiated magnetic flux and energy. Moreover, this result for the sudden turn-on of current became the basis for extending the calculation to other electric current profiles, starting with a linear current ramp of finite duration and then continuing to curved ramps and a ramp that approached its maximum value asymptotically. The corresponding turn-off of current was a decaying exponential like one might encounter when turning a voltage off and watching the current decay due to the resistance in a wire. Each calculation provided an idea for another calculation, and the results kept coming. Because I was documenting these results and their derivations as I went along, a lengthy manuscript was evolving.

Some results, like the way that an outgoing bubble of flux tears away from a dipole field, seemed to be a basic, underlying theme in the radiation process, and reminded me of the detachment that sometimes occurs when a coronal mass ejection leaves the Sun's corona. Other results, like the calculation of energies and fluxes, seemed to be details of the specific current profiles and geometries. But even they had general properties, like the way that energy was distributed between the field and the radiation, and the fact that the bubble from a sudden turn-off would escape with all of the energy that had been stored in the field. The use of space-time maps to describe the field evolution was a new twist in the analysis, and revealed a fascinating behavior, that was remarkably similar to electron scattering via electron-positron pair production, as seen in a Feynman diagram (Feynman 1949, 1985).

This book is the documentation of all these results. It is an in-depth study of a specific problem and shows how far one can go with a simple idea. I have tried to illustrate the results with sequences of maps of the changing magnetic fields and with space-time maps along the "equatorial" radius. These time-lapse sequences are taken from longer and higher cadence movies, which I hope to provide for the online version of the book. But, at the same time, I have tried to keep the book self-contained as a matter of convenience, so that the reader will be able to learn everything from the figures without having to go to the online movies.

This book is a reference that I can turn to when I want to refresh my mind on how some aspect of the radiation process works in a vacuum. It may be useful for other people who have already learned about magnetic fields in advanced courses on electromagnetism and for those who want to improve their understanding with a different approach. Regardless of whether or not the book is a step toward understanding transient magnetic fields in a conducting medium like the Sun's corona, we have not lost anything by learning what happens in a vacuum. And the comparison may help in both areas.

It is a pleasure to acknowledge Hannah Kaufman at Springer for editorial advice in the preparation of this manuscript. I am grateful to my wife, Marybeth, for putting up with this activity for such a long time. What started as a simple calculation to find out how magnetic fields change with time turned out to be a lot more work than I anticipated, and I am not sure that I would do it again. But I am glad to have this documentation and the understanding that I obtained by creating it. I hope that others will enjoy reading the book and find it useful for their own understanding of transient magnetic fields.

Neil R. Sheeley, Jr.
Alexandria, VA, USA
September 13, 2019

Contents

Chapter 1
Introduction

Most of us learned about magnetic fields as children playing with bar magnets, horseshoe magnets, ceramic magnets from the refrigerator door, or electromagnets obtained by connecting a coil of wire to the terminals of a dry cell or battery. We imagined that magnets are surrounded by field lines similar to the patterns of iron filings around a bar magnet or to the meridional lines of the Earth's field that we inferred from northward-pointing compass needles. In elementary courses in electromagnetism, we learned more about these field lines, including how to draw them. Then, in advanced courses, we learned that the field lines do not really exist after all because they appear different to observers moving relative to each other and may disappear entirely for some observers.

But these field lines become real again when we look at the atmosphere of the Sun, whose magnetically aligned loop structures are unmistakable. Magnetic concentrations dot the Sun like cities on a map, and its plasma structures are the "iron filings" that trace out field lines. Unlike cities, which are fixed on the map, the magnetic concentrations on the Sun move and their field line patterns continually adjust. When opposite magnetic poles come together, the loops of plasma that join them collapse and disappear. When opposite poles move apart, the loops expand. Sometimes, expanding loops become disconnected from the Sun and escape into space, like large bubbles of radiation from an antenna.

How well do we understand the expansion and contraction of magnetic field lines, even in the absence of the highly conducting plasma that surrounds the Sun? We are generally familiar with the collapse that accompanies the ohmic dissipation of the current in a wire. But how do the infinitesimal decreases of current become known to a distant field point? Also, what happens if the current is turned off suddenly, or turned on suddenly? It is easy to imagine that waves carry this information outward to the field point at the speed of light, but it is more difficult to imagine the detailed properties of those waves. This requires a time-dependent calculation.

N. R. Sheeley, Jr., *Transient Magnetic Fields*,
https://doi.org/10.1007/978-3-030-40264-8_1

In principle, we could begin with the magnetic field produced by a steady electric current in a long, straight wire. At each point in space, this field has an "azimuthal" direction relative to the wire, and we imagine a volume of space filled with concentric field lines circulating around the wire. Now, what happens when the current is turned off?

If the current had been produced by a voltage applied to the ends of the wire, the turn-off might be accomplished by suddenly disconnecting the voltage and letting the current decay gradually by ohmic dissipation in the wire. We can imagine the concentric field lines collapsing toward the wire and meeting their outward-diffusing counterparts at the surface. In this case, we would expect the collapse to occur on the diffusion time scale, $\tau \sim \mu_0 \sigma R^2$, where $\mu_0 = 4\pi \times 10^{-7}\,\mathrm{Hm}^{-1}$ is the so-called vacuum permeability ($4\pi \times 10^{-7}\,\Omega\mathrm{sm}^{-1}$ for readers who have forgotten what henries are), σ is the conductivity of the wire, and R is its radius. For a copper wire, $\sigma \approx 6 \times 10^7\,\Omega^{-1}\mathrm{m}^{-1}$. If its cross-sectional radius is $R = 0.25$ mm, then the diffusion time scale, τ, would be about 4.7 μs. If that seems like a short time, remember that light moves at the speed $c = 3 \times 10^8\,\mathrm{ms}^{-1}$ or about a foot per nanosecond. Thus, while the field is diffusing through the copper wire of 0.25 mm radius, a light wave would travel about 1.4 km. If such a wave were released when the voltage was turned off, how would it relate to the much slower collapse of loops surrounding the wire?

Also, what would happen if the current (rather than the voltage) were turned off suddenly? Presumably, a wave would move out from the wire at the speed of light, carrying knowledge of the turn-off with it. For a field point far from the wire, the information would not arrive for a while, and its field would remain unchanged during this time. We can imagine that when the wave finally did arrive, there would be a short interval of complicated adjustment after which the field strength would drop to zero. If the current were turned down to a lower value, rather than off, then presumably a similar process would occur, leaving a field strength corresponding to the final value of the current.

What would these waves look like and how would they fit into the magnetic field before and after the transient? I wanted to find the answers to these questions, but for a more realistic model than the current in an infinitely long, straight wire. Although the long, straight wire gives a particularly simple magnetic field configuration, it has some drawbacks. For example, how can the current be turned on suddenly if the wire is very long? Similarly, how can a field point be located far from a wire whose length presumably exceeds all other dimensions of the problem? To avoid these questions, I wanted to find a model in which the current is confined to a finite volume of space, but which still gives a relatively simple magnetic field configuration. As we will see in the next chapter, the current source of a uniformly magnetized sphere provides such a model.

To describe this model, I will use the coordinate system shown in Fig. 1.1. It is the conventional spherical coordinate system (r, θ, ϕ), where r is the radial distance, θ is the polar angle, and ϕ is the azimuthal angle directed counterclockwise from the x-axis. I will often refer to these coordinates in terms of a latitudinal-longitudinal coordinate system on a sphere like the Sun or Earth. In this case, the z-axis is the

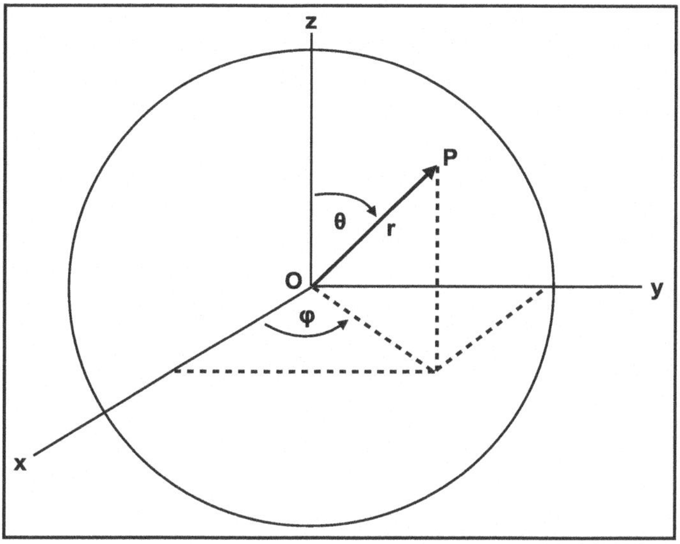

Fig. 1.1 The spherical coordinate system used in this book

polar axis of the sphere, θ is co-latitude, and ϕ is longitude. Occasionally, I will refer to the circle where $r = R$ and $\theta = \pi/2$ as the equator and call the xy-plane the equatorial plane. Likewise, the points where $\theta = 0$ and $\theta = \pi$ will sometimes be called the north and south poles, respectively, when they lie on the sphere.

For a current source that is symmetric about the z-axis, the magnetic field will be independent of ϕ and have components B_r and B_θ, which point outward along r and downward in the direction of increasing θ, respectively. This means that when I plot the magnetic field lines in the yz-plane, positive values of B_θ will correspond to a meridional field with a downward component, and negative values will correspond to a meridional field with an upward component along the z-axis.

The MKS Standard International (SI) system of units will be used throughout the book. This means that μ_0 (the vacuum permeability equal to $4\pi \times 10^{-7}\,\text{Hm}^{-1}$) and ϵ_0 (the vacuum permittivity approximately equal to $8.854 \times 10^{-12}\,\text{Fm}^{-1}$) will appear frequently. In particular, energies of the magnetic and electric fields will be of the form $B^2/2\mu_0$ and $\epsilon_0 E^2/2$, respectively, where B and E are the magnetic and electric field strengths.

Radial distances, r, will usually be expressed in units of R, the radius of the sphere, and times will be expressed in units of R/c, the light transit time along the radius of the sphere. In fact, I will refer time to the start of the current ramp, and use the symbol, r_L, referring to the distance that the leading edge of the disturbance has gone in units of R since the current was turned on. When necessary, I will modify the computed expressions to give the unnormalized dimensions, simply by including an extra R or R/c. There are two cases where these insertions must be done with care: the first case occurs when we study forces. In this case, we must choose between a unit given by magnetic pressure times cross-sectional area, $(B_0^2/2\mu_0)(\pi R^2)$ (where B_0 is the uniform magnetic field strength in the sphere), or by volume energy of the field per unit radius, $(B_0^2/2\mu_0)(4\pi R^3/3)/R$. The difference between these expressions corresponds to a factor of 3/4. The second case occurs when we consider the fields of two concentric spheres of different radii. In this case, we must remember which radius is being used as the reference and express the other radius in terms of that reference radius. With these definitions and alerts, the reader should be well prepared for the discussions that follow.

Chapter 2
The Vector Potential for a Current Step

We begin with a sphere of radius R with no electric current on its surface and no magnetic field in or around it. Then, we turn on a current at the surface of the sphere. Consider an idealized scenario in which someone at the center of the sphere suddenly strikes a match at time $t = 0$. The light from the match moves outward at speed c, and arrives at the sphere at the time $t = R/c$.

As soon as the uniformly spaced receptors on the surface see this light, they turn on the current at their individual locations, producing the current density, **J**, given by

$$\mathbf{J}(r', \theta', \phi') = \left(\frac{3}{2}\right) \frac{B_0}{\mu_0} \sin\theta' \, \delta(r' - R) \, \mathbf{e}_{\phi'}. \tag{2.1}$$

Here, B_0 is the uniform field strength that eventually occurs inside the sphere, (r', θ', ϕ') are spherical polar coordinates in the system whose polar axis is parallel to this uniform field, δ is the Dirac delta function restricting the current to the surface of the sphere, and μ_0 is the vacuum permeability, used in the MKS system of units. The unit vector $\mathbf{e}_{\phi'}$ indicates that the current circulates azimuthally in the ϕ' direction, and the $\sin\theta'$ factor reflects the latitude variation of current density, which is largest at the equator where $\theta' = \pi/2$ and vanishes at the poles where $\theta' = 0$ and π. The total current on the sphere is $\int\int J_{\phi'} r' dr' d\theta' = 3B_0 R/\mu_0$.

We have used primed coordinates to distinguish the location of the current from the location of the distant point (r, θ, ϕ) from which we are making *in situ* observations of the magnetic and electric fields. At this field point, contributions from this instantaneous current step are arriving at progressively later times from correspondingly more distant locations on the sphere. We can keep track of this process using the well-known solution of the wave equation for the vector potential $\mathbf{A}(\mathbf{r}, t)$ (Panofsky and Phillips 1962)

N. R. Sheeley, Jr., *Transient Magnetic Fields*,
https://doi.org/10.1007/978-3-030-40264-8_2

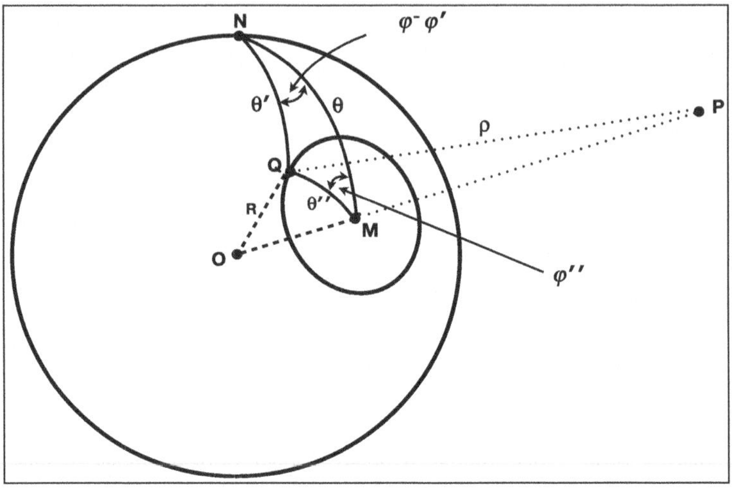

Fig. 2.1 Spherical geometry used to derive the vector potential from the current source. The field point, $P(r, \theta, \phi)$, sees contributions from points, $Q(R, \theta'', \phi'')$, on the circle centered at the sub-field point M. As time progresses, this circle expands to include contributions from more distant current elements. Dotted lines lie outside the sphere, solid lines lie on the sphere, and dashed lines lie inside

$$\mathbf{A}(\mathbf{r}, t) = \int_{V'} \frac{\mu_0 \, [\mathbf{J}(\mathbf{r}', t')]}{4\pi |\mathbf{r} - \mathbf{r}'|} \, dV', \tag{2.2}$$

where V' indicates that the integration is to be taken over the volume occupied by the current, and the brackets indicate that the current is to be evaluated at the "retarded" time t' at the current source, not at the time t at the field point. The boldfaced $\mathbf{r}$ and $\mathbf{r}'$ refer to the field point and the current source, whose coordinates are (r, θ, ϕ) and (r', θ', ϕ'), respectively.

Next, we evaluate this integral using the geometry shown in Fig. 2.1. The field point, $P(r, \theta, \phi)$, receives contributions from the spherical surface at later times from more distant locations. Information from the nearest point, M, comes first and is followed by contributions from points, Q, on circles of increasingly larger angular radius, θ''. The current elements at point Q on each circle are given by Eq. (2.1) in terms of the singly primed variables relative to the "north" pole N. Thus, in terms of the primed coordinates, the equation for $\mathbf{A}(\mathbf{r}, t)$ becomes

$$\mathbf{A}(\mathbf{r}, t) = \left(\frac{3}{8\pi}\right) B_0 R^2 \int\int \frac{\sin\theta' \, \mathbf{e}'_\phi}{|\mathbf{r} - \mathbf{r}'|} \sin\theta' d\theta' d\phi'. \tag{2.3}$$

In this equation, the range of integration is complicated when expressed in terms of the singly primed coordinates, but it is relatively simple when expressed in terms of the doubly primed coordinates (θ'', ϕ'') whose pole lies at the sub-field point M.

In particular, the range of ϕ''-integration is $[0, 2\pi]$ and the range of θ''-integration is $[0, \theta''_{max}]$, where θ''_{max} is the polar angle of the source circle that contributes to the field at time t. Therefore, we need to convert from the singly primed coordinates (θ', ϕ') to the doubly primed coordinates (θ'', ϕ'') to obtain these relatively simple ranges of integration. Although this is a somewhat tedious exercise, it leads to a relatively simple result that will be the basis for most of our subsequent discussions.

We begin by expressing the unit vector $\mathbf{e}_{\phi'}$ in terms of the unprimed unit vectors $(\mathbf{e_r}, \mathbf{e}_{\theta}, \mathbf{e}_{\phi})$. This seemingly formidable task can be done in a straightforward manner by expressing each of the spherical coordinate unit vectors $(\mathbf{e_r}, \mathbf{e}_{\theta}, \mathbf{e}_{\phi})$ in terms of the rectangular coordinate unit vectors $(\mathbf{e_x}, \mathbf{e_y}, \mathbf{e_z})$ and then taking the scalar product of these expressions for $(\mathbf{e_r}, \mathbf{e}_{\theta}, \mathbf{e}_{\phi})$ with the analogous expression for $\mathbf{e}_{\phi'}$, obtained by replacing the unprimed variables in the $\mathbf{e}_{\phi}$-equation by singly primed variables. Combining the resulting three components of $\mathbf{e}_{\phi'}$, we obtain

$$\mathbf{e}_{\phi'} = \sin\theta \ \sin(\phi - \phi')\ \mathbf{e_r} + \cos\theta \ \sin(\phi - \phi')\ \mathbf{e}_{\theta} + \cos(\phi - \phi')\ \mathbf{e}_{\phi}. \tag{2.4}$$

Next, we note that the integral in Eq. (2.3) contains $\mathbf{e}_{\phi'}$ as part of the factor $\sin\theta'\mathbf{e}_{\phi'}$. Referring to Eq. (2.4), we see that this combination contains the factors $\sin\theta' \sin(\phi - \phi')$ and $\sin\theta' \cos(\phi - \phi')$. Fortunately, these factors can be obtained from the spherical triangle NMQ in Fig. 2.1:

$$\sin\theta' \ \sin(\phi - \phi') = \sin\theta'' \ \sin\phi'' \tag{2.5a}$$

$$\sin\theta' \ \cos(\phi - \phi') = \sin\theta \ \cos\theta'' - \cos\theta \ \sin\theta'' \ \cos\phi''. \tag{2.5b}$$

Here, Eq. (2.5a) is obtained by applying the law of sines, and Eq. (2.5b) is obtained by applying the law of cosines twice (once to the angle $\phi - \phi'$ and once to the angle ϕ'') and combining the results. With these two equations, we can now remove the singly primed variables from $\sin\theta'\mathbf{e}_{\phi'}$.

Next, because the areal element $\sin\theta' d\theta' d\phi'$ is a constant, we can replace it with $\sin\theta'' d\theta'' d\phi''$. This leaves the denominator $|\mathbf{r} - \mathbf{r}'|$ as the only remaining function of primed coordinates. In Fig. 2.1, this distance is indicated by the symbol ρ, which can be expressed in terms of θ'' using the plane triangle POQ.

$$\rho^2 = r^2 + R^2 - 2rR \ \cos\theta''. \tag{2.6}$$

Now, we are in a position to evaluate the integral in Eq. (2.3). Performing the ϕ''-integration over the interval $[0, 2\pi]$, we see that the radial and meridional components vanish, leaving only the factor $\sin\theta \cos\theta''$ from Eq. (2.5b). The result is

$$A_{\phi}(r, \theta, t) = \frac{3}{4} \ B_0 R^2 \ \sin\theta \int_0^{\theta''_{\max}} \frac{\sin\theta'' \ \cos\theta''}{\rho(\theta'')} \ d\theta''. \tag{2.7}$$

Next, we use Eq. (2.6) to change the integral in Eq. (2.7) to a simpler integral over the variable ρ.

$$A_\phi(r,\theta,t) = \frac{B_0}{2}\frac{\sin\theta}{r^2}\frac{3}{4}\int_{\rho_{\min}}^{\rho_{\max}} \left(r^2 + R^2 - \rho^2\right) d\rho, \tag{2.8}$$

where the lower limit, $\rho_{\min}$, is obtained by setting $\theta'' = 0$ in Eq. (2.6). $\rho_{\min} = r - R$ if the field point P lies outside the sphere where $r > R$ and $\rho_{\min} = R - r$ if P lies inside the sphere where $r < R$. The upper limit, $\rho_{\max}$, corresponds to the elapsed time, t, since the most recent information left the sphere, and is given by $\rho_{\max} = ct - R$. Using these limits to evaluate this integral, we obtain

$$A_\phi(r,\theta,r_L) = \frac{B_0 R}{2}\frac{\sin\theta}{r^2} f(r,r_L), \tag{2.9}$$

where the distance r is now expressed in units of the radius R. The time, t, is expressed in units of R/c and is represented by the symbol r_L. We think of r_L as the distance that the disturbance has moved (in units of R) during the time t. The function $f(r, r_L)$ is given by

$$f(r,r_L) = \left[3\left(r^2+1\right)(r_L-1) - (r_L-1)^3 \pm 2\left(r^3-1\right)\right]/4. \tag{2.10}$$

Here, the minus sign applies outside the sphere where $r > 1$ and the plus sign applies inside. In the remainder of our discussion, we shall use $f_o(r, r_L)$ and $f_i(r, r_L)$ to refer to $f(r, r_L)$ in these two domains. In particular, we write

$$f_o(r,r_L) = \left[3\left(r^2+1\right)(r_L-1) - (r_L-1)^3 - 2\left(r^3-1\right)\right]/4 \tag{2.11a}$$

$$f_i(r,r_L) = \left[3\left(r^2+1\right)(r_L-1) - (r_L-1)^3 + 2\left(r^3-1\right)\right]/4, \tag{2.11b}$$

where $f_o(r, r_L)$ and $f_i(r, r_L)$ represent the waves that move outward and inward, respectively, from the newly created current on the sphere.

The function, $f(r, r_L)$, is the final product of our calculation. It is a kind of "reduced vector potential" obtained when the factor, $\sin\theta/r^2$, is removed from the vector potential in Eq. (2.9). The relatively simple algebraic form of this expression is a pleasant surprise considering the somewhat tortuous procedure that led to it. In hindsight, we can show that $f_o(r, r_L)$ and $f_i(r, r_L)$ are valid solutions to our problem by substituting the corresponding values of $A_\phi(r, \theta, r_L)$ directly into the wave equation for the vector potential

$$\frac{1}{r}\frac{\partial^2(rA_\phi)}{\partial r^2} + \frac{1}{r^2}\frac{\partial}{\partial\theta}\left[\frac{1}{\sin\theta}\frac{\partial}{\partial\theta}(A_\phi\sin\theta)\right] - \frac{1}{c^2}\frac{\partial^2 A_\phi}{\partial t^2} = 0, \tag{2.12}$$

which takes the form

$$f'' - \frac{2}{r} f' - \frac{\partial^2 f}{\partial {r_L}^2} = 0, \tag{2.13}$$

when expressed in terms of the reduced vector potential $f(r, r_L)$. Here, the primes refer to partial derivatives with respect to the spatial variable, r (in units of R) and the time, r_L, is expressed in units of R/c. Only a small amount of algebra is required to show that both $f_o(r, r_L)$ and $f_i(r, r_L)$ satisfy this equation (and that these solutions are therefore relativistically correct as we carefully ensured by using retarded time in the geometric approach associated with Fig. 2.1).

We still need to satisfy the boundary conditions for the sudden current step, but that is relatively easy to do. The sharp leading edges of these waves follow directly from the facts that $f_o(r, r) = 0$ and $f_i(r, 2 - r) = 0$, as one can easily see from the factored representations of f_o and f_i:

$$f_o(r, r_L) = (r_L - r)\{2r^2 - (r_L - 3)(r_L + r)\}/4 \tag{2.14a}$$

$$f_i(r, r_L) = (r_L + r - 2)\{2r^2 - (r_L + 1)(r_L - r - 2)\}/4. \tag{2.14b}$$

Also, when the ingoing wave reaches the center, it turns around and moves outward (or passes through the center and moves outward toward the other side) as a wave of the form $-f_i(-r, r_L)$. The first minus sign ensures that the composite field vanishes at the center of the sphere and the second minus sign produces an outgoing wave. This wave adds to the fields that have been left behind the original ingoing and outgoing waves, giving $f_i(r, r_L) - f_i(-r, r_L) = r^3$ inside the sphere and $f_o(r, r_L) - f_i(-r, r_L) = 1$ outside the sphere. As we shall see next, these values correspond to the vector potential of the dipole field of a uniformly magnetized sphere. Also, for a uniform field of strength, B_0, the jump of B_θ/μ_0 across the surface is $(3/2)(B_0/\mu_0)\sin\theta$, which equals the current, $(3/2)(B_0/\mu_0)\sin\theta$, flowing azimuthally on the surface.

Chapter 3
Fields and Field Lines

3.1 Basic Field Equations

The electric and magnetic fields are easily derived from the vector potential using the relations

$$\mathbf{E}(\mathbf{r}, t) = -\frac{\partial \mathbf{A}}{\partial t} \tag{3.1a}$$

$$\mathbf{B}(\mathbf{r}, t) = \nabla \mathbf{x} \mathbf{A}. \tag{3.1b}$$

In the spherical coordinate system (r, θ, ϕ), the non-vanishing components of the axisymmetric field become

$$B_r = \left(\frac{1}{R}\right) \frac{1}{r \sin\theta} \frac{\partial}{\partial \theta} (A_\phi \sin\theta), \tag{3.2a}$$

$$B_\theta = -\left(\frac{1}{R}\right) \frac{1}{r} \frac{\partial}{\partial r} (r A_\phi), \tag{3.2b}$$

$$E_\phi = -\left(\frac{c}{R}\right) \frac{\partial A_\phi}{\partial r_L}, \tag{3.2c}$$

where r is radial distance in units of R, and $r_L = ct/R$. Substituting A_ϕ from Eq. (2.9), we obtain

$$B_r = B_0 \, r^{-3} \, f \, \cos\theta, \tag{3.3a}$$

Electronic Supplementary Material The online version of this chapter (https://doi.org/10.1007/978-3-030-40264-8_3) contains supplementary material, which is available to authorized users.

N. R. Sheeley, Jr., *Transient Magnetic Fields*,
https://doi.org/10.1007/978-3-030-40264-8_3

$$B_\theta = \frac{1}{2}\, B_0\, r^{-3} \left(f - r\, \frac{\partial f}{\partial r} \right) \sin\theta, \tag{3.3b}$$

$$E_\phi = -\frac{1}{2}\, c B_0\, r^{-2} \left(\frac{\partial f}{\partial r_L} \right) \sin\theta, \tag{3.3c}$$

where $f(r, r_L)$ is the reduced vector potential given by Eq. (2.10).

From these equations, we can infer several properties of the reduced vector potential $f(r, r_L)$. First, the continuity of B_r ensures that f is also continuous across all spatial boundaries. Second, the discontinuity of B_θ at the surface of the sphere implies that f' is also discontinuous there. Third, the electric field, E_ϕ, exists only when the vector potential changes with time (i.e., when $\partial f/\partial r_L \neq 0$).

Also, it is easy to see that the f-values of r^3 and 1 that were left behind the waves in the previous section correspond to the dipole field extending from a uniformly magnetized sphere. Substituting $f = r^3$ into Eqs. (3.3), we obtain the uniform field of strength B_0 inside the sphere:

$$B_r = B_0 \cos\theta, \tag{3.4a}$$

$$B_\theta = -B_0 \sin\theta, \tag{3.4b}$$

$$E_\phi = 0. \tag{3.4c}$$

By substituting $f = 1$ into Eqs. (3.3), we obtain the field components of the corresponding static dipole:

$$B_r = B_0\, r^{-3} \cos\theta, \tag{3.5a}$$

$$B_\theta = \frac{1}{2}\, B_0\, r^{-3} \sin\theta, \tag{3.5b}$$

$$E_\phi = 0. \tag{3.5c}$$

For this field, the jump of B_θ/μ_0 at $r = 1$ is $(3/2)(B_0/\mu_0)\sin\theta$, which equals the magnitude of the current flowing azimuthally on the surface of the sphere, as mentioned above.

3.2 Magnetic Field Lines

A useful way of seeing the transient variation of the field is through the evolving pattern of magnetic field lines. For the axisymmetric field of an azimuthal current source, the field line passing through the point (r, θ, ϕ) is uniquely determined by r_L and the magnetic flux, Φ, passing outward through the circle of radius $r \sin\theta$ centered on the polar axis. Taking B_r from Eq. (3.3a), we can express the flux as

$$\Phi = \int_0^{\theta} B_r 2\pi r^2 \sin\theta d\theta = \Phi_0 \frac{f(r, r_L)}{r} \sin^2\theta, \qquad (3.6)$$

where $\Phi_0 = \pi R^2 B_0$ is the total flux trapped in the sphere. Thus, at a given time, r_L, the field lines are obtained by plotting contours of constant $\Phi/\Phi_0 = \{f(r, r_L)/r\} \sin^2\theta$. We will use red contours for field lines that circulate clockwise in the right half plane where $y > 0$. Likewise, we will use red contours for their symmetric counterparts that circulate counter clockwise in the left half plane where $y < 0$. We will use blue contours for field lines that circulate in the opposite direction.

Figure 3.1 shows the evolving magnetic field lines in the yz-plane for a sequence of times from $r_L = 1.3$ to $r_L = 4.0$. Two waves (indicated by dotted circles) diverge from the current immediately after it is turned on at time $r_L = 1.0$. The ingoing wave reaches the center of the sphere at $r_L = 2$, and turns around (or passes through the center and moves outward toward the other side). The net effect is to "fold back on itself" and become the trailing edge of the disturbance whose leading edge now lies 2 radii ahead of it.

As this trailing edge moves out through the sphere, it leaves a uniform field in its wake. At the same time, the equatorial field just outside of the sphere weakens (as

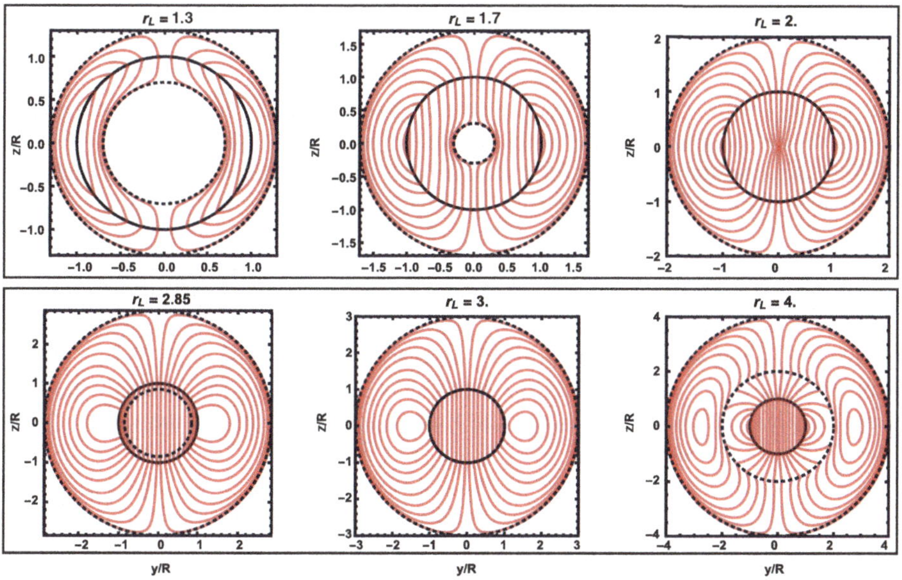

Fig. 3.1 A meridional section of the magnetic field lines produced by a sudden current step at time $r_L = 1$, showing the expulsion of flux from the sphere at $r = 1$. Two wave fronts (dotted circles) diverge from the current. The ingoing wave turns around when it reaches the center at $r_L = 2$, and eventually becomes the trailing boundary of a toroidal bubble of detached flux. The net effect is to leave a uniform field inside the sphere and a dipole field outside. The spatial scale is varied to match the outward motion of the disturbance

indicated by the decreased density of contours at $r_L = 2.85$), and an O-type neutral point forms at $r = 1$. This happens at the time $r_L = 1 + 6^{1/3} \approx 2.817$, obtained by solving $B_\theta(1, r_L) = 0$ (equivalently, by solving $f - rf' = 0$ when $r = 1$). After the trailing edge crosses the surface at $r_L = 3.0$ and moves outward, it pinches off the loops that are draped around the O-type neutral point, forming two components. The upper component becomes part of the loop system that surrounds the outgoing neutral point, and which I shall often call a bubble of flux. The lower component is the latest addition to the stationary loops of the dipole field that is left behind. In a later chapter, we shall use space-time maps to examine this process in more detail.

It is instructive to consider the flux contours of the escaping bubble. We begin by noting that the contour of maximum strength occurs near $r = r_L - 1$, midway between the leading and trailing edges of the bubble. We can see this by setting $\partial\Phi/\partial r = 0$, which is equivalent to $f_o - rf_o' = 0$. Outside the sphere, the flux fraction is given by Eq. (3.6) with $f = f_o(r, r_L)$. Substituting the value of f_o given by Eq. (2.11a), we find that

$$4r^3 - 3(r_L - 1)r^2 - r_L^2(r_L - 3) = 0. \tag{3.7}$$

Then, defining the offset from the center by $x = r_L - 1 - r$ and introducing the variable $a = r_L - 1$ for simplicity, we find

$$4x^3 - 9ax^2 + 6a^2x - (3a + 2) = 0. \tag{3.8}$$

For $x/a \ll 1$, $x \approx 1/(2a)$, which means that $r \approx (r_L - 1) - 1/\{2(r_L - 1)\}$ and that $r \to r_L - 1$ as $r_L \to \infty$. So the peak flux begins slightly behind the center and moves toward the center as the bubble moves away from the sphere.

Next, we consider the field lines themselves. From Eq. (3.6), we know that they are described by contours of constant Φ/Φ_0 given by

$$\frac{\Phi}{\Phi_0} = \frac{f_o(r, r_L)}{r}\sin^2\theta = \frac{1}{4}\frac{(r_L - r)}{r}\left\{2r^2 - (r_L - 3)(r_L + r)\right\}\sin^2\theta. \tag{3.9}$$

In the two-dimensional system, we choose a coordinate system centered in the bubble where $r = r_L - 1$ and the latitude is $\Lambda = 0$, as indicated in the left panel of Fig. 3.2. Then, we rewrite the flux fraction in terms of the radial offset $\delta = r - (r_L - 1)$, and obtain

$$\frac{\Phi}{\Phi_0} = \frac{(1 - \delta)}{4r}\{3(1 + \delta)r + (1 - \delta)(2 + \delta)\}\cos^2\Lambda. \tag{3.10}$$

When $r \gg 1$, this equation reduces to

$$\frac{\Phi}{\Phi_0} \approx \frac{3}{4}\left(1 - \delta^2\right)\cos^2\Lambda. \tag{3.11}$$

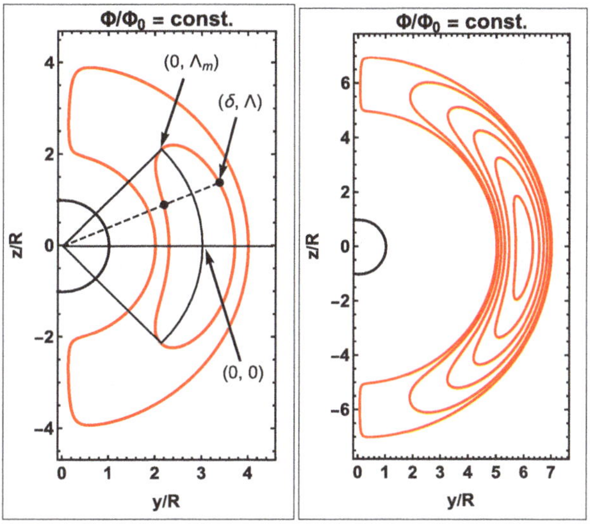

Fig. 3.2 Approximate flux contours for the escaping bubble after the sudden turn-on of current. In this approximation, all of the contours are assumed to have detached from the sphere, which only happens in the limit that $r_L \to \infty$ (*cf.* Fig. 3.1). (Left) Sketch used for deriving the contours with coordinates expressed relative to the equatorial location of the peak field of the bubble; (right) plot of the contours for $r_L = 7$

When $\delta = 0$, Λ has its maximum value, Λ_m, given by

$$\cos^2 \Lambda_m = \frac{4}{3}\frac{\Phi}{\Phi_0}. \tag{3.12}$$

In terms of this parameter, the equation for the flux contours becomes

$$\delta^2 + \frac{\cos^2 \Lambda_m}{\cos^2 \Lambda} = 1, \tag{3.13}$$

and applies when the bubble lies far from the sphere.

From Eq. (3.13), we can see that $\Lambda = \Lambda_m$ when $\delta = 0$. Similarly, when $\Lambda = 0$, δ has its extreme values of $\delta_m = \pm \sin \Lambda_m$. So the red contour is somewhat like a bent ellipse with its semi-minor axis, $\delta_m = \sin \Lambda_m$, and its semi-major axis equal to $(r_L - 1)\Lambda_m$. Thus, for $\Lambda_m \ll 1$, the ratio of these quantities is $r_L - 1$, and these contours become increasingly squashed with time.

Referring to the right panel of Fig. 3.2, we can see that the flux contours form a concentric "nest" with the outer contour corresponding to $\Lambda_m = \pi/2$ and $\Phi = 0$, and the inner contour corresponding to $\Lambda_m = 0$ and $\Phi/\Phi_0 = 3/4$. In this figure, the red curves are contours of $\cos^2 \Lambda_m \approx \{0, 0.1, 0.3, 0.5, 0.7, 0.9\}$, corresponding to contours of Φ/Φ_0 equal to 3/4 of these values. In summary, as r_L becomes large, the escaping bubble forms a nest of closed contours, carrying a total flux, $(3/4)\Phi_0$. As we shall see next, this nest of loops also encloses a displacement current.

The second term on the right side of Maxwell's equation

$$\nabla \times \left(\frac{\mathbf{B}}{\mu_0}\right) = \mathbf{J} + \epsilon_0 \frac{\partial \mathbf{E}}{\partial t} \tag{3.14}$$

is often called the displacement current (or displacement current density), analogous to the real current density, **J**, in the first term. For the fields of the sudden turn-on, **J** lies on the surface of the sphere, and the displacement current is enclosed by the nest of loops in the outgoing bubble. We can probably guess the value of this current, but it is interesting to see what it really is. We can do the calculation either by integrating $\epsilon_0 \partial E_\phi/\partial t$ over the cross-sectional area between the outgoing waves (as $\epsilon_0 \int (\partial E_\phi/\partial t) r dr d\theta$), or by integrating **B** around the edge of this region (as $\oint \mathbf{B}\cdot\mathbf{ds}/\mu_0$). For simplicity, we shall take the former approach. Substituting E_ϕ from Eq. (3.3c), we obtain an exact expression for the current, i_ϕ, as

$$i_\phi(r_L) = -\frac{B_0 R}{2\mu_0}\int_0^\pi \sin\theta d\theta \int_{r_L-2}^{r_L} r^{-1}\frac{\partial^2 f_o(r, r_L)}{\partial r_L^2} dr$$
$$= \frac{3}{2}\frac{B_0 R}{\mu_0}(r_L - 1)\ln\left(\frac{r_L}{r_L - 2}\right), \tag{3.15}$$

where we have inserted the unit of length, R. When the bubble breaks out of the sphere at $r_L = 3$, this current has the value $i_\phi = 3(B_0 R/\mu_0)\ \ln 3$, which is 1.10 times the amount of real current, $3(B_0 R/\mu_0)$, on the surface of the sphere that we obtained by evaluating $\int\int J_{\phi'} r' dr' d\theta'$ in Chap. 2.

Expanding the expression for $i_\phi(r_L)$ as a power series in $r_L - 1$, we obtain

$$i_\phi(r_L) = 3\frac{B_0 R}{\mu_0}\left\{1 + \frac{1}{3}(r_L - 1)^{-2} + \frac{1}{5}(r_L - 1)^{-4} + \cdots\right\}. \tag{3.16}$$

Thus, there is real current of magnitude $3B_0 R/\mu_0$ on the surface of the sphere and a displacement current whose magnitude asymptotically approaches this value in the closed loops of the outgoing bubble. However, there is no current in the static dipole field that fills the space between them.

3.3 Magnetic Flux

We can use Eq. (3.6) to derive the flux in this bubble as a function of time r_L. For $\theta = \pi/2$, Eq. (3.6) gives the meridional flux that crosses the equatorial plane outside the radius r (i.e., $\Phi = 2\pi \int_r^\infty B_\theta(r, \pi/2)\ r dr = \Phi_0\ f(r, r_L)/r$). When r_L lies in the interval (2.82, 3), the flux lying between the edge of the bubble at $r = 1$ and the center of the bubble at r_0 is given by

$$\frac{\Phi}{\Phi_0} = \frac{f_o(r_0, r_L)}{r_0} - f_o(1, r_L), \tag{3.17}$$

where r_0 is obtained by setting $B_\theta(r_0, r_L) = 0$ (equivalently $f_o - r f_o' = 0$ with $r = r_0$). When $r_L = 3$, the trailing edge crosses the spherical surface, leaving space outside the sphere where the legs of the draped field lines pinch together, adding flux to the bubble and closed loops at the top of the growing dipole field.

Consequently, for $r_L > 3$, we use $r_L - 2$ as the trailing edge of the bubble, and evaluate the toroidal flux in the bubble using the equation

$$\frac{\Phi}{\Phi_0} = \frac{f(r_0, r_L)}{r_0} - \frac{f(r_L - 2, r_L)}{r_L - 2} = \frac{f(r_0, r_L)}{r_0} - \frac{1}{r_L - 2}. \tag{3.18}$$

This normalized flux, Φ/Φ_0, increases from 0 at $r_L \approx 2.82$ when the bubble was born, to 1/8 at $r_L = 3$ when the trailing edge crossed the surface. To evaluate this flux when $r_L \gg 1$, we need to expand the series for r_0 to more terms than we used in the previous subsection, and, for comparison with Eq. (3.18), it will be convenient to expand in terms of $r_L - 2$ rather than $r_L - 1$. In this case, the solution to Eq. (3.8) is $x = (1/2)b^{-1} - (1/6)b^{-2} + (5/24)b^{-3}$, where $b = r_L - 2$ and $r_0 = r_L - 1 - x$. Consequently,

$$\frac{f_o(r_0, r_L)}{r_0} = f'(r_0, r_L) = \frac{3}{2} r_0 (r_L - 1 - r_0) = \frac{3}{4} + \frac{1}{2} \frac{1}{r_L - 2}, \tag{3.19}$$

which is the total flux that lies ahead of the center of the bubble at $r = r_0$. Part of this flux is the draped flux, $1/(r_L - 2)$, which we subtract to obtain the bubble flux:

$$\frac{\Phi}{\Phi_0} = \frac{3}{4} - \frac{1}{2} \frac{1}{r_L - 2}. \tag{3.20}$$

Thus, to first order in $(r_L - 2)^{-1}$, the bubble flux increases at one-half the rate at which the draped flux decreases. Because the bubble obtains its flux from the conversion of draped flux, this means that half of the converted flux is being lost at the center of the bubble as closed loops shrink and disappear. The net result is that the draped flux approaches 0 and the bubble flux approaches the value $(3/4)\Phi_0$.

Figure 3.3 provides a graphical display of these flux changes. The dashed orange curve indicates the total flux crossing the surface of the sphere. This total flux does not include flux that circulates around the bubble without intersecting the sphere. Prior to $r_L = 2.82$, there is no bubble, and all of the external flux circulates back through the sphere. This flux is $f_i(1, r_L) = (r_L - 1)[6 - (r_L - 1)^2]/4$, which has a maximum value of $\Phi/\Phi_0 = \sqrt{2} \approx 1.414$ when $r_L = 1 + \sqrt{2} \approx 2.414$. As the trailing wave crosses the sphere, it leaves a uniform static field with r^3 of a unit of flux behind it. This means that the total flux suddenly stops decreasing and the orange curve flattens at $r_L = 3$ when $r = 1$ and $\Phi/\Phi_0 = 1$. From this time onward, the flux crossing the sphere is balanced by the sum of the flux that is draped ahead of the bubble and the growing amount of closed flux in the dipole. The interesting point to remember is that for a while the sphere contains more flux than the single unit that will eventually reside there, which means that there must be a process that removes this excess flux.

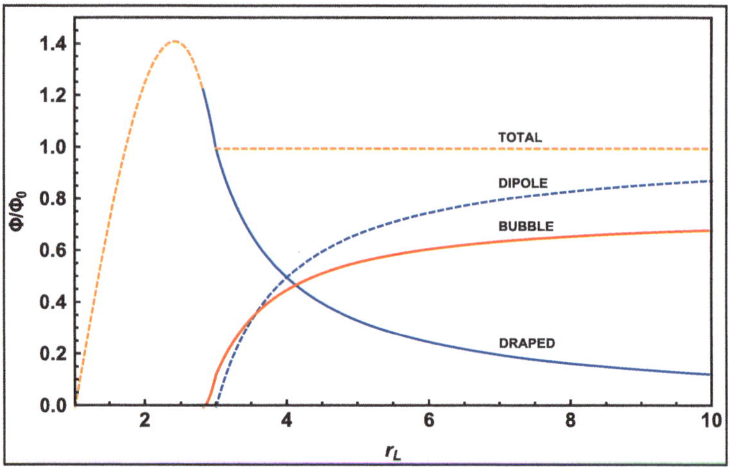

Fig. 3.3 Plots of the flux Φ/Φ_0 outside the sphere as a function of time, showing the conversion of draped flux to dipole flux after $r_L = 3$ and gradual increase of bubble flux toward 3/4 as it moves outward. The plotted quantities include the total flux passing outward across the sphere (dashed orange curve), the parts of this flux that are draped over the bubble (solid blue curve) and left behind as dipole field lines (dashed blue curve), and the toroidal flux within the bubble (solid red curve)

The solid red curve shows the toroidal flux in the bubble. As mentioned above, this flux increases from 0 to 1/8 during the interval [2.82, 3.00]. During this time, there is no dipole field, and all of the flux that crosses the sphere is draped around of the bubble. Therefore, the blue curve and the dashed orange curve coincide during this short interval. After $r_L = 3$, the total flux becomes constant, the draped flux decreases, the dipole flux increases by the same amount, and the bubble flux begins its asymptotic approach to the value 3/4.

Thus, the bubble eventually sweeps up the entire unit of draped flux that lies ahead of it at $r_L = 3$ and leaves an equal amount of dipole flux behind it. This raises an interesting question. If the bubble has 1/8 of a unit at $r_L = 3$ when the total flux is 1 unit, and if the bubble gradually acquires all of this draped flux, then why does the bubble end up with only 3/4 of a unit instead of 9/8 of a unit of flux? Evidently, 3/8 of a unit of toroidal flux was lost as the bubble moved outward.

We can confirm this deduction by calculating the change in the amount of flux that lies ahead of the point r_0 (where $B_\theta = 0$) between $r_L = 3$ (when $r_0 = 3/2$) and a much greater time, r_L. This change is given by

$$\frac{\Delta\Phi}{\Phi_0} = \frac{f_o(r_0, r_L)}{r_0} - \frac{f_o(3/2, 3)}{3/2} = \frac{f_o(r_0, r_L)}{r_0} - \frac{9}{8}. \tag{3.21}$$

As before, r_0 is determined from $f - rf' = 0$. Letting $r_L \to \infty$, we find that $r_0 \to r_L - 1$ and $f_o(r_0, r_L)/r_0 \to 3/4$. Consequently, $\Delta\Phi/\Phi_0 = -3/8$, in agreement with the amount of flux that we expected to have been lost. Also, during this time, the radial distance, r_0, of the neutral point (or azimuthal ring) moved forward by 0.5 of a radius from $r_L - 1.5$ to $r_L - 1$, and the neutral point became centered in the bubble.

We note that the reconnection point (or ring) behind the bubble is not a neutral point where $B_\theta = 0$, like the O-type neutral point within the bubble itself. Instead, it is a place where B_θ is discontinuous like the one that occurs at the leading edge of the bubble. At both of these locations, $c\Delta B_\theta = \Delta E_\phi$, analogous to the jump, ΔB_θ, that occurs at the spherical current sheet for the static dipole field.

This illustrates that for transient fields, additional discontinuities of B_θ occur at moving boundaries where E_ϕ (and therefore $\partial f/\partial r_L$) is discontinuous, and that care must be used in the analysis of such moving boundaries. In particular, these two kinds of discontinuity may occur simultaneously when a moving boundary sweeps past a current sheet, as occurred at $r = 1$ in this example when $r_L = 3$ and the trailing edge of the bubble moved past the sphere.

In this regard, it is interesting to examine the flux that lies beyond the center, r_0, of the bubble defined by $B_\theta(r_0, r_L) = 0$ (equivalently $f - rf' = 0$). The flux that originates in the sphere and is draped over the bubble (i.e., the draped flux) is given by

$$\frac{\Phi_{dr}}{\Phi_0} = \begin{cases} f_o(1, r_L), & 2.82 < r_L < 3 \\ 1/(r_L - 1), & r_L > 3 \end{cases} \tag{3.22}$$

Although this draped flux is continuous when $r_L = 3$, its time derivative is not continuous. The flux that has detached from the sphere (i.e., the bubble flux) and lies ahead of the center contains the same continuity-braking terms:

$$\frac{\Phi_{bub}}{\Phi_0} = \begin{cases} f_o(r_0, r_L)/r_0 - f_o(1, r_L), & 2.82 < r_L < 3 \\ f_o(r_0, r_L)/r_0 - 1/(r_L - 1), & r_L > 3 \end{cases} \tag{3.23}$$

Consequently, the time derivative of this bubble flux is also discontinuous when $r_L = 3$, as shown by the slight discontinuity in the slope of the red curve in Fig. 3.3 (The corresponding discontinuity of the slope of the solid blue curve is less visible in that figure.) However, the terms that produce this slope discontinuity cancel out in the sum of the draped and bubble flux, giving a continuous function, Φ_{comb}, with continuous derivatives:

$$\frac{\Phi_{comb}}{\Phi_0} = f_o(r_0, r_L)/r_0, \qquad r_L > 2.82. \tag{3.24}$$

We can calculate the time derivative of this combined (draped plus bubble) flux using

$$\frac{d}{dr_L}\left[\frac{\Phi_{comb}}{\Phi_0}\right] = \frac{d}{dr_L}\left[\frac{f_o(r_0, r_L)}{r_0}\right] = \frac{1}{r_0}\frac{\partial f_o}{\partial r_L} + \frac{1}{r_0^2}\frac{\partial r_0}{\partial r_L}\left(r_0\frac{\partial f_o}{\partial r} - f_o\right), \tag{3.25}$$

where $r_0 = r_0(r_L)$ is the evolving radius of the O-type null point, obtained by solving

$$f_o(r, r_L) - r\,\frac{\partial f_o(r, r_L)}{\partial r} = 0 \tag{3.26}$$

for r. For this value of r_0, the last term in Eq. (3.25) vanishes and we obtain

$$\frac{d}{dr_L}\left[\frac{\Phi_{comb}}{\Phi_0}\right] = \frac{1}{r_0}\frac{\partial f_o}{\partial r_L}. \tag{3.27}$$

Combining this equation with Eq. (3.3c), we obtain

$$\frac{d\Phi_{comb}}{dt} = -2\pi r_0\, E_\phi, \tag{3.28}$$

where we have returned to unnormalized units and used $\Phi_0 = \pi R^2 B_0$. This is a familiar form of Lenz's Law applied to the circular ring through the O-type neutral points. It states that the electric potential around that ring is equal to the time rate of change of flux through (and returning outside) that ring.

Chapter 4
Radiated Energy and the Sudden Turn-Off

Not only does the bubble carry magnetic flux away from the sphere, but it also carries energy. Let us evaluate the energy that flows across a sphere of radius r during the passage of the bubble. The flow rate is given by the Poynting vector, $\mathbf{E} \times \mathbf{B}/\mu_0$, whose radial component $S_r = -E_\phi B_\theta/\mu_0$.

We wish to evaluate the integral of S_r over the spherical surface of radius, r, during the time that the bubble passes that surface:

$$\mathcal{E} = \frac{R^3}{c} \int_{r_{L1}}^{r_{L2}} dr_L \int_0^{2\pi} d\phi \int_0^{\pi} S_r \, r^2 \sin\theta d\theta, \tag{4.1}$$

where r is radial distance in units of R, and r_L is time in units of R/c with r_{L1} and r_{L2} representing the beginning and end points of the time interval. Using Eqs. (3.3b) and (3.3c) to express the field components in terms of f and its derivatives, we obtain

$$\mathcal{E} = \frac{B_0^2}{2\mu_0} \frac{R^3}{2} \int_0^{2\pi} d\phi \int_0^{\pi} \sin^3\theta d\theta \, r^{-3} \int_{r_{L1}}^{r_{L2}} \frac{\partial f}{\partial r_L} \left(f - r\frac{\partial f}{\partial r} \right) dr_L, \tag{4.2}$$

where $f(r, r_L)$ is given by Eq. (2.10), with the minus sign selected when r lies outside the sphere and the plus sign inside. Obtaining 2π from the ϕ-integration, 4/3 from the θ-integration, and defining an energy unit $\mathcal{E}_0 = (B_0^2/2\mu_0)(4\pi R^3/3)$, we may rewrite Eq. (4.2) as

$$\frac{\mathcal{E}}{\mathcal{E}_0} = \frac{f^2(r, r_{L2}) - f^2(r, r_{L1})}{2r^3} - \frac{1}{r^2} \int_{r_{L1}}^{r_{L2}} \frac{\partial f}{\partial r_L} \frac{\partial f}{\partial r} dr_L. \tag{4.3}$$

Electronic Supplementary Material The online version of this chapter (https://doi.org/10.1007/978-3-030-40264-8_4) contains supplementary material, which is available to authorized users.

N. R. Sheeley, Jr., *Transient Magnetic Fields*,
https://doi.org/10.1007/978-3-030-40264-8_4

Inside the sphere, $r < 1$, the endpoints of the time interval are $r_{L1} = 2 - r$ and $r_{L2} = 2 + r$, and we must use the plus sign in the definition of f. Consequently, $f(r, r_{L1}) = f_i(r, 2 - r) = 0$, $f(r, r_{L2}) = f_i(r, 2 + r) = r^3$, and the first term is $(1/2)r^3$. The second term is $(3/2)r^3$, so $\mathcal{E}/\mathcal{E}_0 = -r^3$. Thus, the energy flows inward across the surface at radius r by just the amount required for a uniform field inside the sphere. In particular, setting $r = 1$, we obtain $\mathcal{E}_0$ as the total energy left inside the sphere of radius R.

Outside the sphere, $r > 1$, the endpoints are $r_{L1} = r$ and $r_{L2} = r + 2$, and we must use the minus sign in the definition of f. Thus, $f(r, r_{L1}) = f_o(r, r) = 0$, $f(r, r_{L2}) = f_o(r, r + 2) = 1$, and the first term is $1/2r^3$. This is the dipole field energy (in units of $\mathcal{E}_0$) that will ultimately be deposited outside the radial distance r. When $r = 1$, it is 1/2, the total energy flowing into the dipole field. The second term is 3/2, independent of r. This constant term dominates the flow at large distances, which means that the bubble escapes with 3/2 of a unit of energy.

In summary, the total energy flowing outward across the surface of a sphere of radius r is given by

$$\frac{\mathcal{E}}{\mathcal{E}_0} = \begin{cases} -r^3, & r < 1 \\ 1/(2r^3) + 3/2, & r > 1. \end{cases} \tag{4.4}$$

Thus, in turning the current on, we cause 3 units of energy to flow away from the current: 1 unit flows inward to provide the volume energy of the uniform field, and 2 units flow outward to provide the volume energy of the dipole field (1/2 a unit) and the radiated energy that is carried away by the bubble (3/2 of a unit).

We can evaluate this energy source directly by remembering that, in general, the Poynting flux and field energy are balanced by a volume source term, $-\mathbf{J}\cdot\mathbf{E}$, according to the equation

$$-\mathbf{J}\cdot\mathbf{E} = \nabla\cdot\left(\frac{\mathbf{E}\times\mathbf{B}}{\mu_0}\right) + \frac{\partial}{\partial t}\left(\frac{\epsilon_0 E^2}{2} + \frac{B^2}{2\mu_0}\right), \tag{4.5}$$

where ϵ_0 and μ_0 are the vacuum permittivity and vacuum permeability, respectively, used in the MKS system of units. In our case, $\mathbf{J}$ is the current density given by Eq. (2.1) and $\mathbf{E}$ is the electric field of the transient. Thus, the total energy supplied by the source is

$$\mathcal{E} = \int_0^\infty r^2 dr \int_0^\pi \sin\theta d\theta \int_0^{2\pi} d\phi \int_1^3 (-J_\phi E_\phi) dr_L, \tag{4.6}$$

where $J_\phi = (3/2)(B_0/\mu_0)\,\delta(r-1)\sin\theta$ and $E_\phi = -(1/2)(cB_0)\,r^{-2}\,(\partial f/\partial r_L)\sin\theta$, as given by Eqs. (2.10) and (3.3c), respectively. The θ-integral gives 4/3, the ϕ-integral gives 2π, the δ-function in the r-integral requires that we set $r = 1$, and the r_L-integral gives $f(1, 3) - f(1, 1) = 1$. In terms of the energy unit,

$\mathcal{E}_0 = (B_0^2/2\mu_0)(4\pi R^3/3)$, we obtain the expected result that $\mathcal{E}/\mathcal{E}_0 = 3$. Thus, by suddenly turning the current on, we set up a spherical wave whose electric field acts on the current for the short interval $2R/c$ and extracts 3 units of energy from us as we produce the current.

Now, what would happen if the current were suddenly turned off? Because Maxwell's equations are linear, the resulting fields would be the sum of the old field left behind the turn-on bubble and the new field produced by suddenly adding a current going in the opposite azimuthal direction around the sphere. This sudden turn-off would produce another transient with an electric field acting for a time $2R/c$. However, this electric field would not have an electric current to act on, and therefore could not exchange any more energy. The source term in Eq. (4.5) would vanish, and the divergence of the Poynting flux would be balanced against the time rate of change of the volume energy in the electric and magnetic fields.

Thus, we can express the reduced vector potential as $f_{on}(r, r_L) - f(r, r_L)$, where $f_{on}(r, rL)$ represents the fields left behind the turn-on bubble (r^3 inside the sphere and 1 outside the sphere) and $f(r, r_L)$ is given by Eq. (2.10). In this case, Eq. (4.3) gives

$$\frac{\mathcal{E}}{\mathcal{E}_0} = \begin{cases} +r^3, & r < 1 \\ -1/(2r^3) + 3/2, & r > 1. \end{cases} \tag{4.7}$$

The field energy again flows toward the sphere, but is intercepted by the bubble which carries it away. The net result of turning the current suddenly on and then suddenly off is to radiate all of the energy away and leave no field behind.

Figure 4.1 shows the evolution of the magnetic field lines during the sudden turn-off. The ingoing wave (whose front is indicated by a dotted circle) overpowers the uniform field near the equator. This creates an "evacuated" region with a neutral point where the original red field begins to separate into a small region inside the sphere and a larger region outside the sphere. When the separation is complete, the blue field of reversed polarity (created by turning the current off) becomes visible. As time advances, this reversed-polarity field expands, eroding the red field left in the sphere and sweeping up the remnant of the original dipole field outside the sphere. Eventually, no field lines are left behind the outgoing disturbance either inside or outside the sphere. Thus, the net result of turning the current suddenly on and then (after a pause of at least $2R/c$) suddenly off again is the creation of two bubbles, a red bubble of flux circulating clockwise in the right half plane where $y > 0$ (and counterclockwise in the left half plane) and a blue bubble of reversed-polarity flux circulating in the opposite direction. Each bubble carries away 3/4 a unit of flux ($\Phi_0 = \pi R^2 B_0$) and 3/2 a unit of energy ($\mathcal{E}_0 = (B_0^2/2\mu_0)(4\pi R^3/3)$). All of this energy was provided when the current was turned on. The sudden turn-off was "free"; the blue bubble of reversed-polarity flux acquired its energy from the field created when the current was turned on.

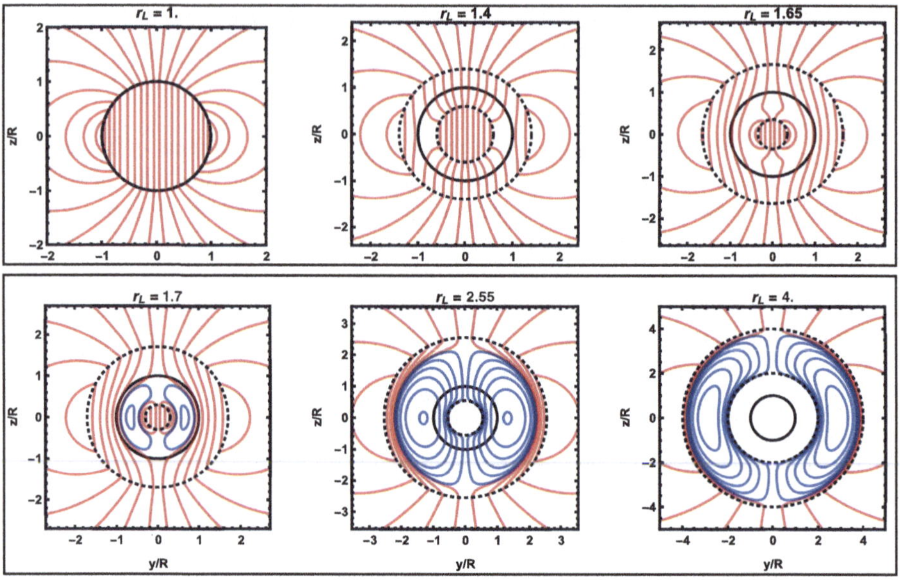

Fig. 4.1 The transient produced by turning the current off suddenly at time $r_L = 1$. Ingoing and outgoing wave fronts are indicated by dotted circles. The ingoing wave creates an "evacuated" region where the field lines pinch together, separating the original field into interior and exterior branches. When the separation is complete, the reverse-polarity (blue) bubble becomes visible as it sweeps away the remnants of the original field. The spatial scale is varied to follow the leading edge of the outgoing disturbance

One of the original motives of this study was to understand what happens to the field surrounding a current when the current is turned off. Do the field lines collapse or not? Here, we have seen that if the current is turned off suddenly and the observer is located far from the current, the field lines do not collapse. The observer does not know that the current has been turned off until the bubble arrives with that information. By this time, it is too late; there are no field lines left to collapse. In the next chapter, we shall see that field lines sometimes do collapse when the field is turned off slowly over an extended interval of time.

Chapter 5
The Fields of a Linear Current Ramp

We have seen that when the current given by Eq. (2.1) is turned on suddenly, a bubble of flux moves outward leaving the dipole field of a uniformly magnetized sphere in its wake. We can use this result to evaluate the field when the current is turned on gradually. The idea is to regard the gradual increase as a sequence of infinitesimal sudden increases, like small steps in a long staircase. Because Maxwell's equations are linear, we can combine the individual vector potentials of these current steps to obtain the vector potential for the entire ramp.

To illustrate this procedure, we consider a ramp in which the current increases linearly from 0 to a final value given by Eq. (2.1) during the ramp time Δt. In keeping with the dimensionless units of the previous chapters, we express this ramp time in units of R/c and call it Δr. If we think of the ramp as a "staircase" of infinitesimal sudden increases , then each "step," dr'_L, will contribute a normalized vector potential, $f(r, r'_L)/\Delta r$, where r'_L corresponds to the leading edge of each infinitesimal bubble. The entire staircase produces a sequence of bubbles (and trailing dipoles) whose leading edges, r'_L, range from r_L to $r_L - \Delta r$ and whose trailing edges, $r'_L - 2$, lie 2 radii behind. The sequence of disturbances then spans the spatial range $(r_L - \Delta r - 2,\ r_L)$.

The resulting normalized vector potential, $F(r, r_L)$, may then be obtained by adding together the contributions of the individual bubbles in the ramp:

$$F(r, r_L) = \frac{1}{\Delta r} \int_{r_L - \Delta r}^{r_L} f(r, r'_L)\, dr'_L. \tag{5.1}$$

Even for this simple linear ramp, the detailed solution is fairly complicated. It depends on whether we are looking inside or outside the sphere, and whether the ramp is large ($\Delta r > 2$) or small ($\Delta r < 2$). In addition, each of these solutions has

---

Electronic Supplementary Material The online version of this chapter (https://doi.org/10.1007/978-3-030-40264-8_5) contains supplementary material, which is available to authorized users.

N. R. Sheeley, Jr., *Transient Magnetic Fields*,
https://doi.org/10.1007/978-3-030-40264-8_5

three parts corresponding to a central region and its leading and trailing sheaths. Despite these complications, we will keep track of these details not only to learn more about the field of a current ramp, but also to gain insight that will be helpful for describing the fields of other current variations later in this analysis.

Suppose that r lies outside the sphere. If it lies ahead of the pack of bubbles where $r > r_L$, then none of the bubbles will have arrived there yet, and therefore $F(r, r_L) = 0$. If r lies behind the pack where $r < r_L - \Delta r - 2$, then all of the bubbles will have passed by, leaving a dipole field with $F(r, r_L) = 1$. When r lies within the pack, the result depends on whether $\Delta r \geq 2$ or $\Delta r \leq 2$.

Consider the case for which $\Delta r \geq 2$. If r lies in the sheath, $(r_L - 2, r_L)$, near the leading edge of the pack, then the only bubbles that have reached it are those whose leading edges lie between r and r_L. Because their trailing edges have not yet passed r, these bubbles are still making time-dependent contributions rather than the static dipole contributions that follow them. In this case, the f-value, $F_{1o}(r, r_L)$, in this leading sheath is given by

$$F_{1o}(r, r_L) = \frac{1}{\Delta r} \int_r^{r_L} f_o(r, r'_L)\, dr'_L, \tag{5.2}$$

where $f_o(r, r'_L)$ is obtained from Eq. (2.11a). If r lies in the middle region $(r_L - \Delta r, r_L - 2)$, then it receives the maximum number of time-dependent contributions (i.e., the number that can occur in an interval 2 units wide) plus the dipole contributions of additional bubbles that have passed by. Consequently, in this middle region, the f-value, $F_{mo}(r, r_L)$, is given by

$$F_{mo}(r, r_L) = \frac{1}{\Delta r} \int_r^{r+2} f_o(r, r'_L)\, dr'_L + \frac{1}{\Delta r} \int_{r+2}^{r_L} dr'_L = \frac{r_L - 1}{\Delta r}, \tag{5.3}$$

where the first term represents the "saturated" contribution of the bubbles and the second term represents the remaining dipole contributions. Their sum, $(r_L - 1)/\Delta r$, corresponds to the linear growth of a dipole magnetic field with time and a static azimuthal electric field.

Finally, if r lies in the sheath, $(r_L - \Delta r - 2, r_L - \Delta r)$, near the trailing edge of the pack, then the only bubbles still making active contributions are those with leading edges between $r_L - \Delta r$ and $r + 2$; the remainder are providing static dipole contributions. Consequently, the f-value, $F_{2o}(r, r_L)$, in this trailing sheath is given by

$$F_{2o}(r, r_L) = \frac{1}{\Delta r} \int_{r_L - \Delta r}^{r+2} f_o(r, r'_L)\, dr'_L + \frac{1}{\Delta r} \int_{r+2}^{r_L} dr'_L, \tag{5.4}$$

where the first term contains the contributions from the passing bubbles and the second term contains contributions from the static dipoles left behind the older bubbles. Taken together, these three equations provide the complete f-value for the field outside the sphere when $\Delta r \geq 2$. It is easy to see that this combination is continuous for r ranging from $r_L - \Delta r - 2$ to r_L and gives 0 and 1 at the beginning and end of the interval, respectively.

The situation is slightly more complicated inside the sphere because the ingoing waves turn around at the center and move outward over the same region they passed heading inward. The trick is to remember that a given field point, r, continues to feel the time-dependent contribution of an ingoing wave until that wave passes it heading outward, leaving its uniform-field contribution. Let us consider the case for which r_L lies in the range (2, 3), so that the leading edge of the sequence of waves is headed back out through the sphere. If r lies behind the leading edge where $r < r_L - 2$, then it receives active contributions from all of the waves that have passed r on their inward trip, but have not yet reached r on their outward return (i.e., for which r'_L lies in the range $(2 - r, 2 + r)$). Consequently, the f-value, $F_{mi}(r, r_L)$, for this middle region inside the sphere is given by

$$F_{mi}(r, r_L) = \frac{1}{\Delta r}\int_{2-r}^{2+r} f_i(r, r'_L)dr'_L + \frac{1}{\Delta r}\int_{2+r}^{r_L} r^3 dr'_L = \left(\frac{r_L - 1}{\Delta r}\right) r^3, \tag{5.5}$$

where $f_i(r, r'_L)$ is obtained from Eq. (2.11b). For this linear ramp, $F_{mi}(r, r_L)$ takes on the simple form $r^3\ (r_L - 1)/\Delta r$, corresponding to the linear increase of a uniform magnetic field with time and a static azimuthal electric field.

In this manner, one may derive the f-values, $F_{1i}(r, r_L)$ and $F_{2i}(r, r_L)$ for the other cases with $r \leq 1$, as well as for the cases with $\Delta r \leq 2$. We summarize the results for $\Delta r \geq 2$ in the next equations. Inside the sphere, the f-values are

$$\begin{aligned}
F_{bi} &= 0,\\
F_{1i} &= \frac{1}{\Delta r}\int_{2-r}^{r_L} f_i(r, r'_L)dr'_L,\\
F_{mi} &= \frac{1}{\Delta r}\int_{2-r}^{2+r} f_i(r, r'_L)dr'_L + \frac{1}{\Delta r}\int_{2+r}^{r_L} r^3 dr'_L\\
F_{2i} &= \frac{1}{\Delta r}\int_{r_L-\Delta r}^{2+r} f_i(r, r'_L)dr'_L + \frac{1}{\Delta r}\int_{2+r}^{r_L} r^3 dr'_L,\\
F_{ei} &= \frac{1}{\Delta r}\int_{r_L-\Delta r}^{r_L} r^3 dr'_L = r^3,
\end{aligned} \tag{5.6}$$

where F_{bi} and F_{ei} are the beginning and ending values. Outside the sphere, the f-values are

$$
\begin{aligned}
F_{bo} &= 0, \\
F_{1o} &= \frac{1}{\Delta r}\int_{r}^{r_L} f_o(r, r_L')dr_L', \\
F_{mo} &= \frac{1}{\Delta r}\int_{r}^{r+2} f_o(r, r_L')dr_L' + \frac{1}{\Delta r}\int_{r+2}^{r_L} dr_L', \\
F_{2o} &= \frac{1}{\Delta r}\int_{r_L-\Delta r}^{r+2} f_o(r, r_L')dr_L' + \frac{1}{\Delta r}\int_{r+2}^{r_L} dr_L', \\
F_{eo} &= \frac{1}{\Delta r}\int_{r_L-\Delta r}^{r_L} dr_L' = 1,
\end{aligned}
\tag{5.7}
$$

where F_{bo} and F_{eo} are the beginning and ending values. We shall return to these equations later when we wish to obtain f-values and field lines for current ramps with non-linear temporal profiles.

Meanwhile, we will use the expressions for $f_i(r, r_L)$ and $f_o(r, r_L)$ given by Eqs. (2.11a) and (2.11b) to evaluate the integrals in Eqs. (5.6) and (5.7). Inside the sphere, the result is

$$
\begin{aligned}
F_{bi} &= 0, \\
F_{1i} &= \frac{(r + r_L - 2)^2}{16\Delta r}[3(r - r_L + 2)^2 + 8(r - r_L + 2)(r_L - 1) \\
&\quad + 4(r_L - 1)(r_L - 2)], \\
F_{mi} &= \left(\frac{r_L - 1}{\Delta r}\right) r^3, \\
F_{2i} &= r^3 - \frac{(r - r_L + \Delta r + 2)^2}{16\Delta r}[3(r + r_L - \Delta r - 2)^2 - 8(r + r_L - \Delta r - 2) \\
&\quad \times (r_L - \Delta r - 1) + 4(r_L - \Delta r - 1)(r_L - \Delta r - 2)], \\
F_{ei} &= r^3,
\end{aligned}
\tag{5.8}
$$

and outside the result is

$$
\begin{aligned}
F_{bo} &= 0, \\
F_{1o} &= \frac{(r - r_L)^2}{16\Delta r}[3(r - r_L + 2)^2 + 4(r - r_L + 2)(r_L - 1) + 4(r_L - 1)], \\
F_{mo} &= \left(\frac{r_L - 1}{\Delta r}\right),
\end{aligned}
\tag{5.9}
$$

$$F_{2o} = 1 - \frac{(r - r_L + \Delta r + 2)^2}{16\Delta r}[3(r - r_L + \Delta r)^2 + 4(r - r_L + \Delta r)$$
$$\times (r_L - \Delta r - 1) - 4(r_L - \Delta r - 1)],$$
$$F_{eo} = 1.$$

Figure 5.1 provides a graphic display of the regions whose f-values are given in these equations. The left side of this figure shows the transient variation at the start of the current ramp, and the right side shows the transient variation at the end of the ramp. The regions whose f-values are given above are separated by four wave fronts, indicated by dashed lines. Think of these four waves as providing f-values that add to the f-values in the regions ahead of them. In this case, the two waves, U_a and U_b, at the start of the ramp can be represented by the equations:

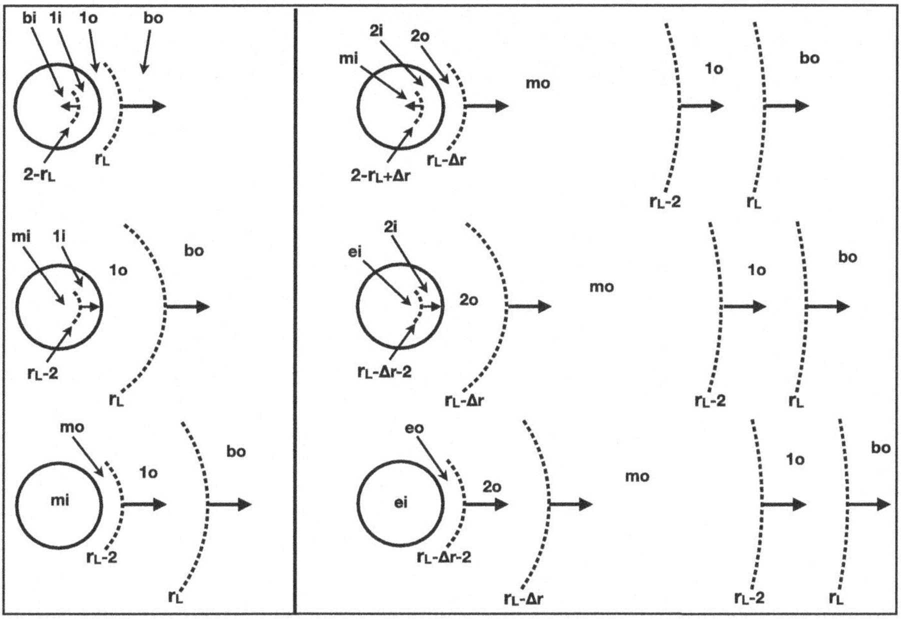

Fig. 5.1 The regions produced by a linear current ramp of length Δr. At the start of the ramp (left panels), ingoing and outgoing waves, indicated by dashed lines, emerge from the surface of the sphere. The ingoing wave turns around at the center and becomes a second outgoing wave located 2 radii behind the first outgoing wave. At the end of the ramp (right panels), a similar pair of ingoing and outgoing waves emerges; again the ingoing wave turns around and follows 2 radii behind the outgoing wave. These four waves divide the outside into beginning (bo) and ending (eo) regions separated by a large middle region (mo) with leading ($1o$) and trailing ($2o$) sheaths. Similar transient regions ($1i$, mi, and $2i$) are created inside the sphere, changing the beginning (bi) into the ending (ei) region. The f-values for these regions are listed in Eqs. (5.6)–(5.9)

$$
\begin{aligned}
U_a^{out}(r, r_L) &= F_{1o} - F_{bo} = F_{1o}, \\
U_b^{in}(r, r_L) &= F_{1i} - F_{bi} = F_{1i}, \\
U_b^{out}(r, r_L) &= -U_b^{in}(-r, r_L) = -F_{1i}(-r, r_L).
\end{aligned} \tag{5.10}
$$

We set $F_{bi} = 0$ and $F_{bo} = 0$ because the initial fields vanish. Likewise, the two waves, V_a and V_b, at the end of the ramp can be represented by

$$
\begin{aligned}
V_a^{out}(r, r_L) &= F_{2o} - F_{mo}, \\
V_b^{in}(r, r_L) &= F_{2i} - F_{mi}, \\
V_b^{out}(r, r_L) &= -V_b^{in}(-r, r_L) = -F_{2i}(-r, r_L) + F_{mi}(-r, r_L).
\end{aligned} \tag{5.11}
$$

In deriving the outward waves, $U_b^{out}(r, r_L)$ and $V_b^{out}(r, r_L)$, from the inward waves, we changed the sign of the ingoing wave to make the superposition vanish at the origin, and we changed the sign of the spatial argument, r, to reverse the direction of motion. This procedure creates four identities and allows the middle and ending f-values to be calculated directly from the f-values in the sheaths:

$$
\begin{aligned}
F_{mi} &= F_{1i} + U_b^{out} = F_{1i} - F_{1i}(-r, r_L) = \left(\frac{r_L - 1}{\Delta r}\right) r^3, \\
F_{mo} &= F_{1o} + U_b^{out} = F_{1o} - F_{1i}(-r, r_L) = \left(\frac{r_L - 1}{\Delta r}\right), \\
F_{ei} &= F_{2i} + V_b^{out} = F_{2i} - F_{2i}(-r, r_L) + F_{mi}(-r, r_L) = r^3, \\
F_{eo} &= F_{2o} + V_b^{out} = F_{2o} - F_{2i}(-r, r_L) + F_{mi}(-r, r_L) = 1.
\end{aligned} \tag{5.12}
$$

Finally, we note that the expressions in Eqs. (5.6) and (5.7) combine to give the identities

$$F_{2o}(r, r_L) = F_{mo}(r, r_L) - F_{1o}(r, r_L - \Delta r) \tag{5.13a}$$

$$F_{2i}(r, r_L) = F_{mi}(r, r_L) - F_{1i}(r, r_L - \Delta r). \tag{5.13b}$$

We combine these identities with Eqs. (5.10) and (5.11) to obtain a relation between the waves, U and V. In particular, we find that

$$V_a^{out}(r, r_L) = -U_a^{out}(r, r_L - \Delta r) \tag{5.14a}$$

$$V_b^{out}(r, r_L) = -U_b^{out}(r, r_L - \Delta r) \tag{5.14b}$$

$$V_b^{in}(r, r_L) = -U_b^{in}(r, r_L - \Delta r). \tag{5.14c}$$

Thus, at a given point in space, each trailing wave, V, is the negative of the corresponding leading wave, U, when it was previously passing through that point.

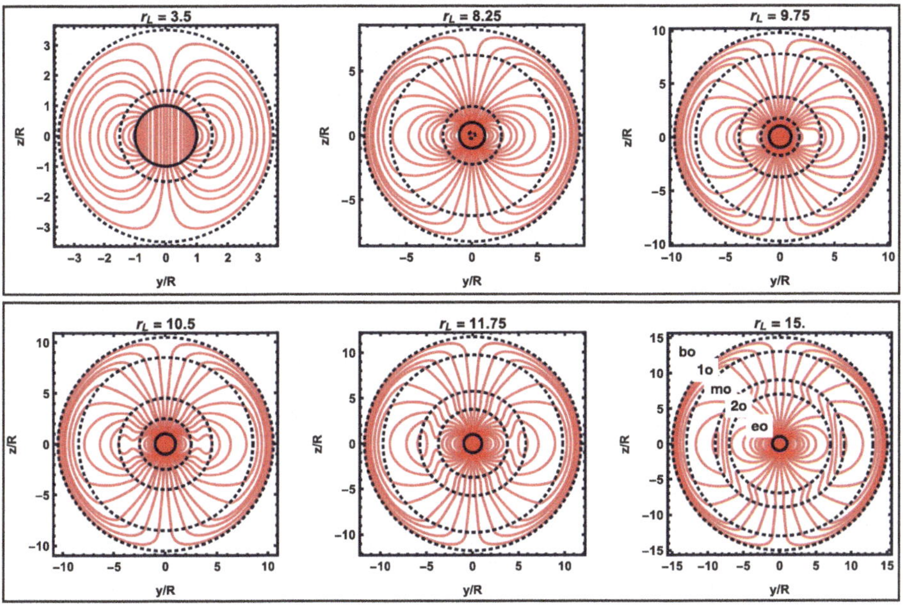

Fig. 5.2 Field lines for a current on-ramp of length $\Delta r = 6$, showing the evolution of the field during the interval $r_L = 3.5$–15. Wave fronts at each end of the ramp (dotted circles) separate a middle region (mo) of growing dipole field lines from sheaths of concentrated flux at its leading ($1o$) and trailing ($2o$) ends. The field-free region at the beginning (bo) and the static dipole at the end (eo) are also indicated. The negative field in the trailing sheath eventually overpowers the positive field of the dipole, producing an "evacuated" region where the legs of stretched loops pinch together and disconnect the outgoing bubble from the static dipole field. The spatial scale is varied to follow the leading edge

Expressed another way, each trailing wave is the negative of the corresponding leading wave, if one allows for the time lag, Δr, between them.

We can combine the f-value Eqs. (5.8) and (5.9) with Eq. (3.6) to plot the field lines (lines of constant flux) for a linearly increasing current ramp of width Δr. Figure 5.2 shows snapshots of the field lines during the interval $r_L = 3.5$–15.0 when $\Delta r = 6$. The dotted circles show the wave fronts that emerged from the sphere at $r_L = 1$ and then again at $r_L = 1 + \Delta r = 7$. By $r_L = 3.5$, the original ingoing wave has reached the center, reversed its direction, and has moved back out through the sphere where it is visible at $r = 1.5$, as the trailing boundary of the leading sheath ($1o$). At this time, the leading sheath dominates the field of view and is filled with flux from the rising current ramp.

By $r_L = 8.25$, the field in this region is becoming compressed into a strong poloidal component, whose $2R$ width seems smaller by comparison with its width in the first panel because we have scaled the images to the distance of the leading edge. Visible at $r = 2.25$, the third wave has left the sphere and forms the trailing boundary of the middle region (mo). The fourth wave is marginally visible inside the sphere at $r = 0.25R$. The middle region (mo) contains a dipole field. As seen in Eq. (5.9), the f-value in this region is $F_{mo} = (r_L - 1)/\Delta r$, which means that the amplitude of this dipole field grows linearly with time. In the next chapter, we will learn that each flux contour (field line) moves linearly outward at its own particular constant speed with the more distant field lines moving fastest and the trailing field lines moving slowest.

The next three panels with $r_L = 9.75$–11.75 show the detachment of this bubble from the static dipole left behind. First, an equatorial region "depleted" of field lines forms in the trailing sheath when $r_L = 9.75$. Then, the legs of the adjacent field lines begin to pinch together when $r_L = 10.5$. By $r_L = 11.75$, the legs have pinched off, forming detached loops with their O-type centers toward the outer edge of this trailing sheath and X-type points where the pinching continues toward the inner edge. Later, at $r_L = 15.0$, the individual regions are well developed, showing the middle region bounded by the compressed fields in its leading and trailing sheaths and the static dipole left behind.

The trailing sheath of strong upward-directed field is responsible for the bubble's separation from the dipole field of the sphere. Because this field strength varies as $1/r_L$, it eventually overpowers the downward-directed dipole field in the middle region (whose strength varies as $1/r_L^2$). This weakens the net field, producing a neutral point (or line) where the legs of the expanding loops pinch together to release the bubble. We can derive these field variations by calculating the flux in the central region and in the sheaths. First, in the trailing sheath, we have

$$\begin{aligned}\frac{\Phi}{\Phi_0} &= -\frac{F(r_L - \Delta r, r_L)}{r_L - \Delta r} + \frac{F(r_L - \Delta r - 2, r_L)}{r_L - \Delta r - 2} \\ &= -\frac{(r_L - 1)/\Delta r}{r_L - \Delta r} + \frac{1}{r_L - \Delta r - 2}.\end{aligned} \tag{5.15}$$

For $r_L \gg \Delta r$, this reduces to $\Phi/\Phi_0 \approx -1/\Delta r$.

In three dimensions, this meridional flux is directed normal to the equatorial plane, and is bounded by the circles of radius $r_L - \Delta r - 2$ and $r_L - \Delta r$. The area of this annulus is $4\pi(r_L - \Delta r - 1)$, giving an average flux per unit area (field strength) of $\Phi_0/4\pi r_L \Delta r$ when $r_L \gg \Delta r$. Recalling that $\Phi_0 = \pi R^2 B_0$, and setting $R = 1$ for our dimensionless units, we can express the poloidal field in the trailing sheath, B_{2o}, in terms of the uniform field strength, B_0, in the sphere:

$$B_{2o} = -(B_0/4\Delta r)\,(1/r_L). \tag{5.16}$$

A similar calculation gives the same flux and field strength in the leading sheath (but with opposite sign because the field points down instead of up). However, in this case, the approximation holds for $r_L \gg 2$ rather than $r_L \gg \Delta r$, which is why the field in the leading sheath becomes compressed more quickly than the field in the trailing sheath, as seen in Fig. 5.2.

By comparison, the flux in the middle region is $(1 - 2/\Delta r)(1/r_L)$. This flux approaches zero as r_L approaches ∞, so that all of the flux in this outgoing bubble is eventually concentrated in the leading and trailing sheaths. Dividing the flux in this middle region by the area of the "equatorial" annulus that it intercepts, we obtain the average field strength

$$B_{mo} = (B_0/2\Delta r)\,(1/r_L^2). \tag{5.17}$$

Thus, B_{mo} falls off more rapidly with r_L than B_{2o} does, and a neutral point eventually forms in the trailing sheath.

Figure 5.3 shows what happens when the static dipole that we just created is turned off by ramping the current linearly back to zero during a time $\Delta r = 6$. In this case, the f-values are equal to the values for the uniformly magnetized sphere (r^3 inside and 1 outside) minus the f-values for the turn-on given by Eqs. (5.8) and (5.9). Now, the leading sheath contains the strong field of negative polarity, which causes the Pinchoff to occur in the leading sheath, rather than the trailing

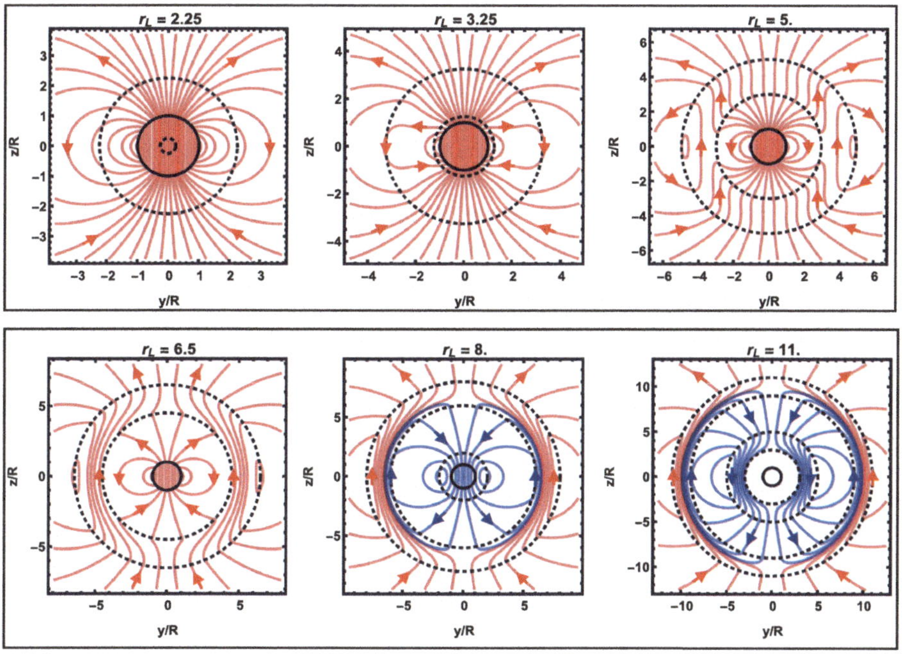

Fig. 5.3 Field lines for a current off-ramp of length $\Delta r = 6$, showing the evolution of the field during the interval $r_L = 2.25$–11.0. Wave fronts at each end of the ramp (dotted circles) outline a middle region bounded by leading and trailing sheaths. The reversed-polarity field in the leading sheath overpowers the initial dipole field and creates an "evacuated region" where the dipole field pinches off. The inward component decays and the outward component is swept away by the blue, reversed-polarity bubble that is now visible dominating the region. The spatial scale is varied to follow the leading edge

sheath. Thus, the stretching, depletion, and reconnection process is visible in the first three panels as the leading sheath forms and moves out from the sphere during $r_L = 2.25$–5.0. Once this reconnection process has started, the legs of the distant dipole field lines move into the X-point near the back of the leading sheath where they pinch together and reconnect. This adds poloidal flux to the sheath and creates shorter dipole field lines. These collapsing field lines cancel their counterparts inside the sphere and weaken its field, as seen in the panel at $r_L = 6.5$. By $r_L = 1+\Delta r = 7$, this cancellation is complete and the field strength inside $r = r_L - 2 = \Delta r - 1 = 5$ has been reduced to zero.

Afterwards, the reversed-polarity field of the turn-off bubble becomes visible, as shown by the blue lines at $r_L = 8.0$. At this time, the leading edge of the trailing sheath is indicated by the dotted circle at $r = 2$ and the back end of this sheath is unobserved at $r = 0$. The leading sheath contains upward-pointing (red) field lines from the back of the original dipole field and upward-pointing (blue) field lines from the front of the reversed-polarity off bubble. By $r_L = 11.0$, the trailing sheath has left the sphere, showing the downward direction of its field and leaving no field behind it. From this point, a growing dipole field of reverse polarity is visible, sandwiched between a leading sheath of upward-pointing field and a trailing sheath of downward-pointing field.

We have seen that the net effect of a linear current ramp of rise time $\Delta r \geq 2$ is to produce a central disturbance of width $\Delta r - 2$ bounded by oppositely directed meridional fields of width 2 (in units of the spherical radius R). As time passes, the strengths of these leading and trailing fields exceed the strength of the meridional component of field in the central region that they are bounding. This is an "edge effect" that we can understand in terms of the contributions from the individual steps of the current ramp.

If we ignore the static dipole contributions after the passage of each bubble, we can represent the superposition of the individual bubbles by uniformly spaced circular loops like those in the top panel of Fig. 5.4. To indicate the clockwise circulation, we split each loop into a leading semi-circle of solid red (a downward-directed field) followed by a trailing semi-circle of dotted blue (an upward-directed field). Because these magnetic loops are spread uniformly, they interfere destructively leaving no net meridional component in the middle region (mo), as shown by their vector sum in the bottom panel.

A radial field component, peaked at the upper and lower boundaries, does exist in the middle region to keep the field solenoidal. It is not difficult to imagine that if we included the individual dipole contributions, they would accumulate to produce a growing dipole field whose strength increases linearly with time along the wave train, analogous to the f-value of $(r_L - 1)/\Delta r$ that we obtained in Eq. (5.9).

However, the important thing to note here is that in the circular regions at the beginning and end of the wave train, the red portion of one loop does not overlap the blue portion of another loop. Consequently, the fields interfere constructively in these regions, producing strong downward and upward meridional components in the two sheaths. These polarities would be reversed if the current were turned off rather than turned on.

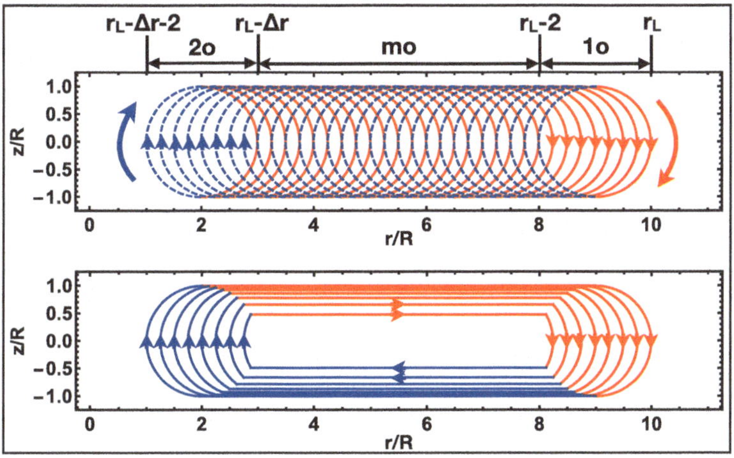

Fig. 5.4 (Top) A series of uniformly spaced circular fields, illustrating how bubbles from the infinitesimal steps of a linear current ramp combine to give sheaths of oppositely directed meridional fields at the leading and trailing ends of the disturbance. In this example, the meridional components add destructively in the middle region, but not at the ends. The result (bottom panel) is a pair of oppositely directed meridional fields in leading ($1o$) and trailing ($2o$) sheaths connected by radial field lines through the middle region (mo)

For completeness, we provide the equations for the fields shown in Fig. 5.4. In the rectangular coordinate system, (r, z), the fields of the individual circles are $(z/R)\mathbf{e}_r - [(r - r_c)/R]\mathbf{e}_z$, where r_c is the r-coordinate of the center of each circle. Also, $\mathbf{e}_r$ and $\mathbf{e}_z$ are unit vectors in the r and z directions. In the middle region, two circles contribute to every field point—the blue part of one and the red part of the other—so that the z-components cancel and the r-components add to give $2(z/R)\mathbf{e}_r$. The normal components of this field are continuous across the boundaries between the three regions, and the jumps of the tangential components are unity (the assumed field strength of the individual loops).

Chapter 6
The Stretching and Pinchoff of Loops

In the previous chapter, we found that a linear on-ramp of electric current produced a growing dipole field sandwiched between sheaths of relatively strong, oppositely directed meridional field. As the structure moved outward, the negative field in the trailing sheath weakened and formed a neutral point where the legs of the stretched field lines could pinch together to separate the outgoing bubble from the static dipole field that was left behind. For a linear off-ramp, the pinchoff occurred in the leading sheath, whose negative field weakened the previously established dipole field and produced a neutral point where the dipole field was split into two parts. The reversed polarity bubble emerged between these parts sweeping them away as it moved inward and then outward from the sphere. In this chapter, we will use space-time maps to examine this process in more detail.

Figure 6.1 shows a space-time map for an on-ramp of width $\Delta r = 3$ together with parts of three field-line maps for this event. The space-time map shows contours of constant flux, $F(r, r_L)/r$, for an equatorial slice of the field lines, plotted as a function of r_L along the horizontal axis and r along the vertical axis. The red tracks that diverge from points along the line $r = 1$ indicate the emergence of flux from the surface of the sphere. For $r < 1$, the tracks eventually line up with a quadratic spacing, indicating the formation of a uniform field inside the sphere. For $r > 1$, the tracks move upward with a positive slope, but eventually they become lined up horizontally with zero slope, representing static loops of the final dipole field outside the sphere.

The interesting part of this space-time map is how the initially steep tracks eventually become horizontal. The first tracks are so steep that they disappear at the edge of the map, but the later tracks reach a maximum height and fall back to their final horizontal positions. In fact, the higher they go, the faster they return, until a point is reached for which the downward return slope is vertical. Beyond this

Electronic Supplementary Material The online version of this chapter (https://doi.org/10.1007/978-3-030-40264-8_6) contains supplementary material, which is available to authorized users.

N. R. Sheeley, Jr., *Transient Magnetic Fields*,
https://doi.org/10.1007/978-3-030-40264-8_6

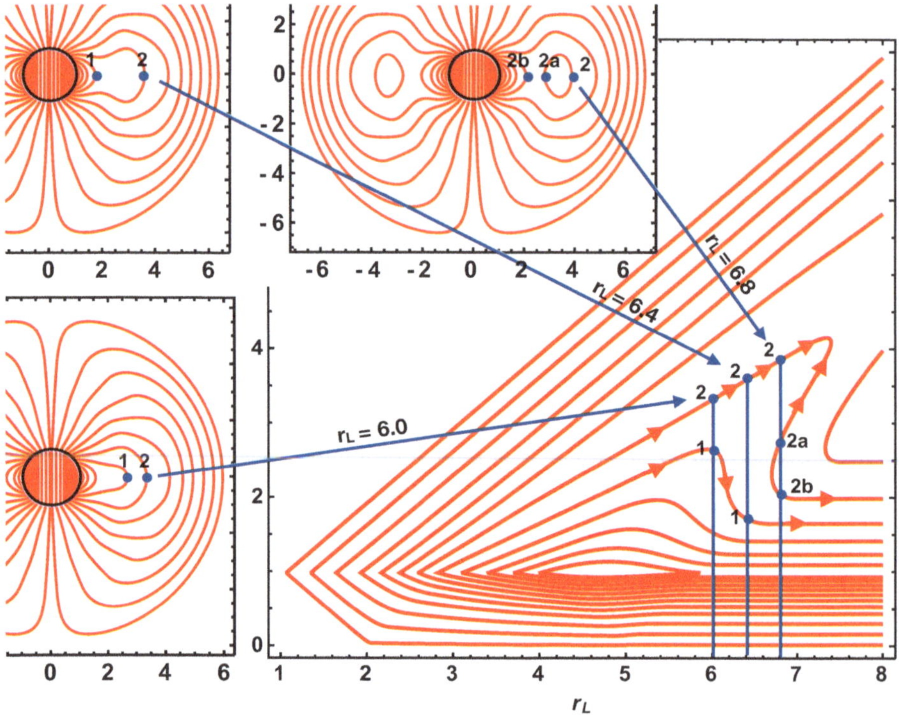

Fig. 6.1 Space-time tracks along the equatorial radius (where $\theta = \pi/2$), showing the radial motion of flux contours (i.e., field lines) during a current on-ramp of length $\Delta r = 3$. The separation and pinchoff of field lines in the magnetic maps correspond to a "switchback" of space-time tracks with flux created at one bend and destroyed at the other. The red arrows indicate the direction of motion and not the orientation of the field lines. The net effect of passing through the switchback is to convert rising flux in the outgoing wave to static flux in the dipole field that is left behind

point, the tracks return, but *via* a pair of space-time switchbacks. Thus, if a point were to move outward along the track, it would return by moving backwards in time—an unusual behavior that requires some explanation.

The three inserts provide this explanation. In the space-time map, the three vertical lines at $r_L = 6.0$, 6.4, and 6.8 correspond to equatorial strips of the fields shown in these inserts. At time $r_L = 6.0$, the field lines labeled #1 and #2 are about $0.5R$ apart. But by $r_L = 6.4$, they are much farther apart; loop #2 has continued upward by about $0.25R$, while loop #1 has fallen by about $1.0R$, giving a total separation of $1.75R$. At $r_L = 6.4$, this sudden fall of point #1 has produced a wide space between the magnetic field lines in the second insert, corresponding to a weakening of the field.

This weakening provides a space where the legs of the stretched field lines pinch together to form a pair of ingoing and outgoing loops indicated by the letters $2a$ and $2b$ in the third insert and on the space-time map at time $r_L = 6.8$. While loop $2b$

moves inward to become part of the static field, loop $2a$ moves outward where it eventually encounters the top of the original loop #2 and disappears with it. Thus, the loops that have risen beyond the critical point never return. Instead, the legs of these stretched loops pinch together to create a pair of oppositely directed loops, one of which moves inward to become part of the static dipole and the other of which moves outward to meet and remove the top of the original loop. In the space-time map, all of these motions are forward in time as indicated by the red arrows.

In this space-time map, the switchbacks that occur after $r_L \approx 6.4$ are the locations of neutral points where loops are either created (i.e., where $2a$ and $2b$ emerged) or destroyed (i.e., where #2 and $2a$ disappeared). These switchbacks occur where the tracks have vertical slopes with $f - rf' = 0$, as one can see from the following argument. First, recall that a single track is defined by a constant flux fraction, $1\Phi/\Phi_0$, and set the derivative of this flux fraction equal to zero as

$$\frac{d}{dr_L}\left(\frac{\Phi}{\Phi_0}\right) = \frac{d}{dr_L}\left[\frac{f(r, r_L)}{r}\right] = \left(\frac{1}{r^2}\right)\left(r\frac{df}{dr_L} - f\frac{dr}{dr_L}\right) = 0. \tag{6.1}$$

Then, use the chain rule for $df(r, r_L)/dr_L$, and write

$$r\left[\frac{\partial f}{\partial r_L} + \frac{\partial f}{\partial r}\frac{dr}{dr_L}\right] - f\frac{dr}{dr_L} = 0. \tag{6.2}$$

Solving for dr/dr_L, we obtain

$$\frac{dr}{dr_L} = \frac{r\,\partial f/\partial r_L}{f - r\,\partial f/\partial r}. \tag{6.3}$$

If we return to unnormalized units and use Eqs. (3.3b) and (3.3c), this relation is equivalent to

$$\frac{dr}{dt} = -\frac{E_\phi}{B_\theta}, \tag{6.4}$$

as the slope of the contours of constant Φ/Φ_0. In particular, the slope is infinite when $B_\theta = 0$ or, equivalently, when

$$f(r, r_L) - r\,\partial f(r, r_L)/\partial r = 0. \tag{6.5}$$

Whereas $f(r, r_L)/r = \Phi/\Phi_0$ is the equation for the contours of constant flux, $f - rf' = 0$ is the equation for the locus of the neutral points of B_θ where the space-time tracks have infinite slopes.

Although $f - r\,\partial f/\partial r = 0$ determines where the tracks are vertical, we also wish to know when, where, and on what track this first happens. This is equivalent to finding the largest flux fraction, Φ/Φ_0, for which $f - r\,\partial f/\partial r = 0$. This happens where $\Phi/\Phi_0 = f(r, r_L)/r = \partial f(r, r_L)/\partial r$ is a maximum. For the linear on-ramp, this maximum occurs within the trailing sheath where $\partial^2 f(r, r_L)/\partial r^2 = 0$. The same equation applies for the linear off-ramp whose maximum occurs inside the leading sheath. However, as we shall see later, for some non-linear current ramps (as well as the sudden turn-on and turn-off), this maximum occurs at a boundary where $\partial^2 f(r, r_L)/\partial r^2$ is discontinuous and not equal to zero. In those cases, we supplement $f - rf' = 0$ with the relevant condition for that boundary (i.e., $r = r_L - 2$ when the point lies on the $1o/mo$ boundary).

The same result is obtained from the field-line maps for the linear on-ramp when we recognize that there are two neutral points, the X-type where loops are created and the O-type where they are destroyed, and we ask where these two neutral points coalesce. This coalescence occurs when both $B_\theta = 0$ and $\partial B_\theta/\partial r = 0$. These conditions imply that $f - r\ \partial f/\partial r = 0$ and $\partial^2 f/\partial r^2 = 0$, provided that the neutral point lies within a sheath where $\partial^2 f/\partial r^2$ is continuous.

Figure 6.2 provides another example, this time for a linear on-ramp of width $\Delta r = 10$. This space-time map shows the transition between the rising loops of a growing dipole field and the static loops of the final field. The long dashed lines indicate the boundaries of the trailing sheath, and the black curve indicates the locus of neutral points where the red curves have infinite slope. The black dot marks the origin of these neutral points at $(r_L, r) = (14.93, 4.05)$ where the black curve itself has infinite slope. This dot separates the curve into a lower segment of X-type neutral points where loops emerge in pairs and an upper segment of O-type neutral points where loops are removed. Note that these neutral points are born together near the middle of the sheath and then separate, moving asymptotically toward opposite boundaries of the sheath.

The flux contours are indicated toward the left and right sides of the figure. It is interesting that the asymptotic values on the right occur at the same locations as the values for the rising contours on the left. This agreement is due to the initial locations at $r_L = 11$, which corresponds to $r_L = \Delta r + 1$ for $\Delta r = 10$. As indicated in Eq. (5.9), the f-values in this middle region are $F_{mo} = (r_L - 1)/\Delta r$, which is 1 when $r_L = \Delta r + 1$. Consequently, the fluxes are distributed as $1/r$ at this location, just as they are on the right side of the figure after the transient has passed by and the final f-value is 1. So in both locations, the flux contours are easily obtained from the radial positions of the respective tracks. In particular, we can see that the maximum flux for the downward return of a loop is $\Phi/\Phi_0 \approx 0.3$, corresponding to beginning and ending locations at $r \approx 3.33$.

Figure 6.3 shows field lines corresponding to the space-time map in Fig. 6.2, and compares the collapse of the 0.3 loop with the pinchoff of the loop with $\Phi/\Phi_0 = 0.27$. As the trailing sheath moves by, the 0.3 loop squeezes slightly and then falls backward, becoming another static dipole loop at $r_L \approx 15.3$ when the back end of the sheath passes by. By comparison, the legs of the 0.27 loop squeeze inward to fill the vacated space. They pinch together at $r_L = 15.4$ to produce a collapsing loop and the lower end of a closed loop that shrinks with time. According to the space-time map in Fig. 6.2, this closed loop disappears when $r_L \approx 15.6$.

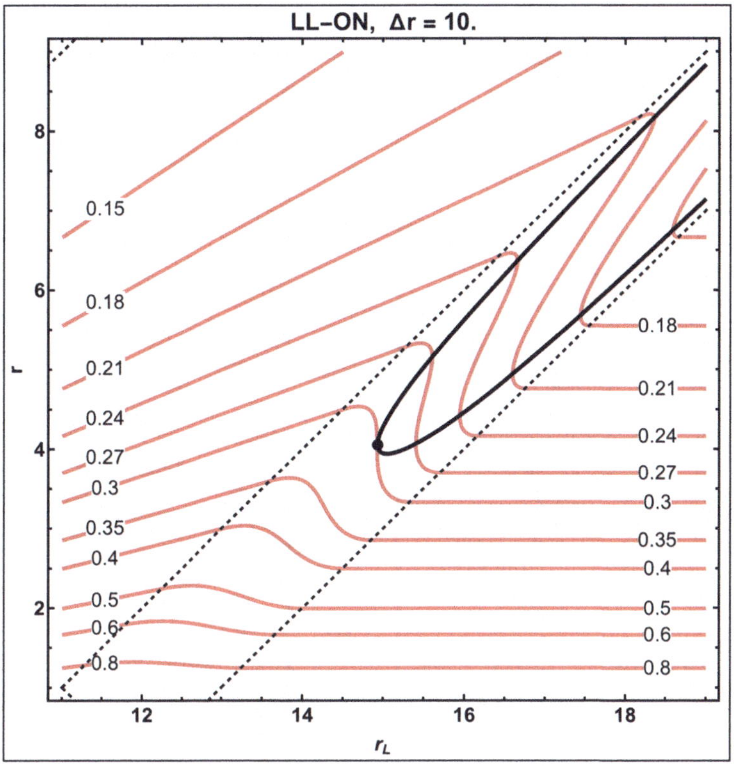

Fig. 6.2 Space-time tracks of constant flux (red), showing the transition between rising loops of the growing dipole field and static loops of the final dipole field. Their tracks bend downward through the trailing sheath (bounded by dashed lines), and some form switchbacks where the slopes are vertical and the field has neutral points (solid black curve). The first of these neutral points (black dot at $(r_L, r) = (14.93, 4.05)$) occurs where the black contour itself has a vertical slope, and indicates the last loop that returns to the dipole. Higher loops with $\Phi/\Phi_0 < 0.30$ are destroyed at the upper branch of the contour and newly created loops return flux to the dipole

Figure 6.4 summarizes the space-time maps for $\Delta r = 6$. Like Figs. 6.1, 6.2, and 6.3, the left panel shows on-ramp behavior, except that the leading sheath is now included. Again, the source of flux moves asymptotically outward along the back of the trailing sheath and the sink moves along the front of the sheath.

The right panel shows the typical patterns for the off-ramp. In this case, the action occurs in the leading sheath where the low-lying loops fall back to the sphere and annihilate their counterparts of uniform field. Beyond the black dot, loops emerge along the lower segment of the neutral-point curve and move in opposite directions to remove flux from the sphere and from the dipole field ahead. At $r_L = 7$, corresponding to $r_L = \Delta r + 1$ in general, all of the positive flux has been removed from the sphere, and the lower branch of the neutral-point curve falls suddenly to $r = 1$.

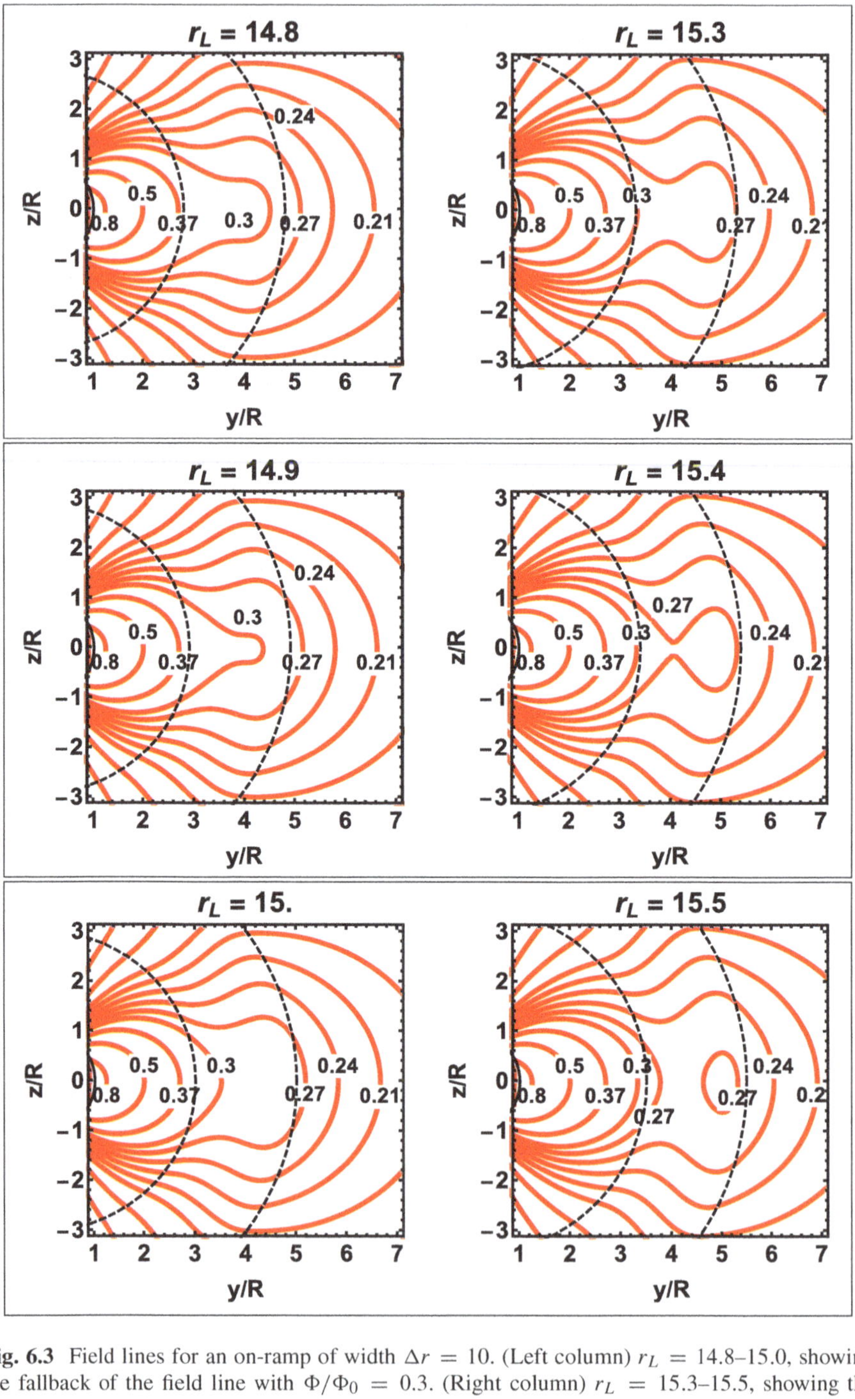

Fig. 6.3 Field lines for an on-ramp of width $\Delta r = 10$. (Left column) $r_L = 14.8$–15.0, showing the fallback of the field line with $\Phi/\Phi_0 = 0.3$. (Right column) $r_L = 15.3$–15.5, showing the separation of the field line with $\Phi/\Phi_0 = 0.27$ into ingoing and outgoing components. The dotted circles indicate the front and back of the trailing sheath ($2o$) as it moves outward across the field

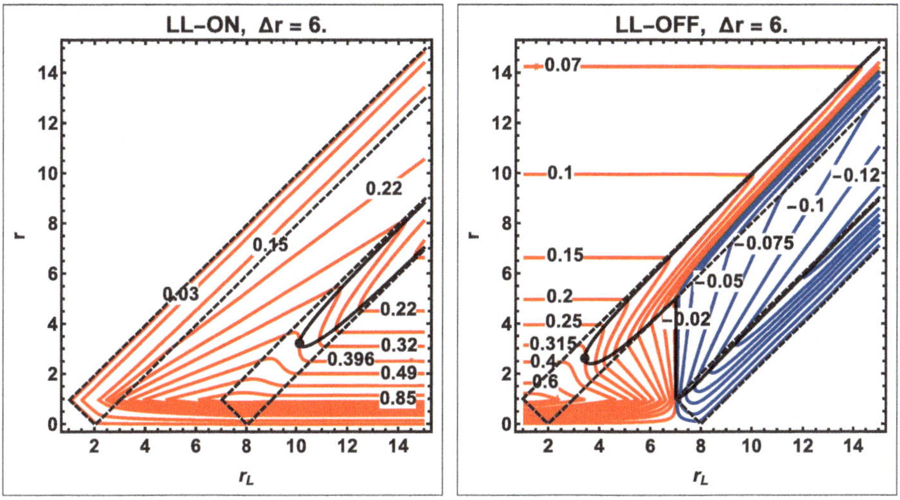

Fig. 6.4 Space-time maps for a linear on-ramp (left) and off-ramp (right), obtained by plotting contours of constant flux $f(r, r_L, \Delta r)/r$ for $\Delta r = 6$. The red and blue tracks show radial motion along the equator (where $\theta = \pi/2$). The dotted boxes outline the leading and trailing sheaths and the black contours indicate the *loci* of neutral points where the tracks have infinite slope. The black dots divide these contours into a lower branch where flux emerges and an upper branch where flux disappears. The red and blue colors refer to the clockwise and counterclockwise circulation of the field lines (when $y > 0$)

We can understand this shift by noting that $f_{mo} = 1 - (r_L - 1)/\Delta r$, obtained by subtracting the on-ramp f-value from the unit f-value of the pre-existing dipole field. Because $f'_{mo} = 0$, this value of f_{mo} is also equal to $f_{mo} - r f'_{mo}$. Consequently, $f_{mo} - r f'_{mo} = 0$ when $r_L = \Delta r + 1$, so that the shift occurs just as the trailing sheath emerges from the sphere.

At this time, the former X-type neutral point is replaced by an O-type neutral point. This O-type neutral point is the source of reversed-polarity flux (blue). This source moves asymptotically outward along the front edge of the trailing sheath, sending negative-polarity flux into this sheath and across the middle region into the leading sheath. The upper segment of the neutral-point curve approaches the front of the leading sheath where its O-type neutral points continue as sinks for the remaining positive flux. The result is an outgoing bubble of negative flux with no field left behind.

Figure 6.5 shows the corresponding field lines during the interval $r_L = 6.4$–9.2 that spans the transition at $r_L = \Delta r + 1 = 7$. The left panels show the pinchoff of the last few red field lines and the corresponding weakening of the field in the sphere. The right column shows the emergence of blue flux at the O-type neutral point located toward the front of the trailing sheath. The inward component of the blue wave reverses direction at the center of the sphere and as this component moves outward, it leaves an evacuated region behind it. Also, the right column shows the continued removal of the original red flux at the front of the leading sheath.

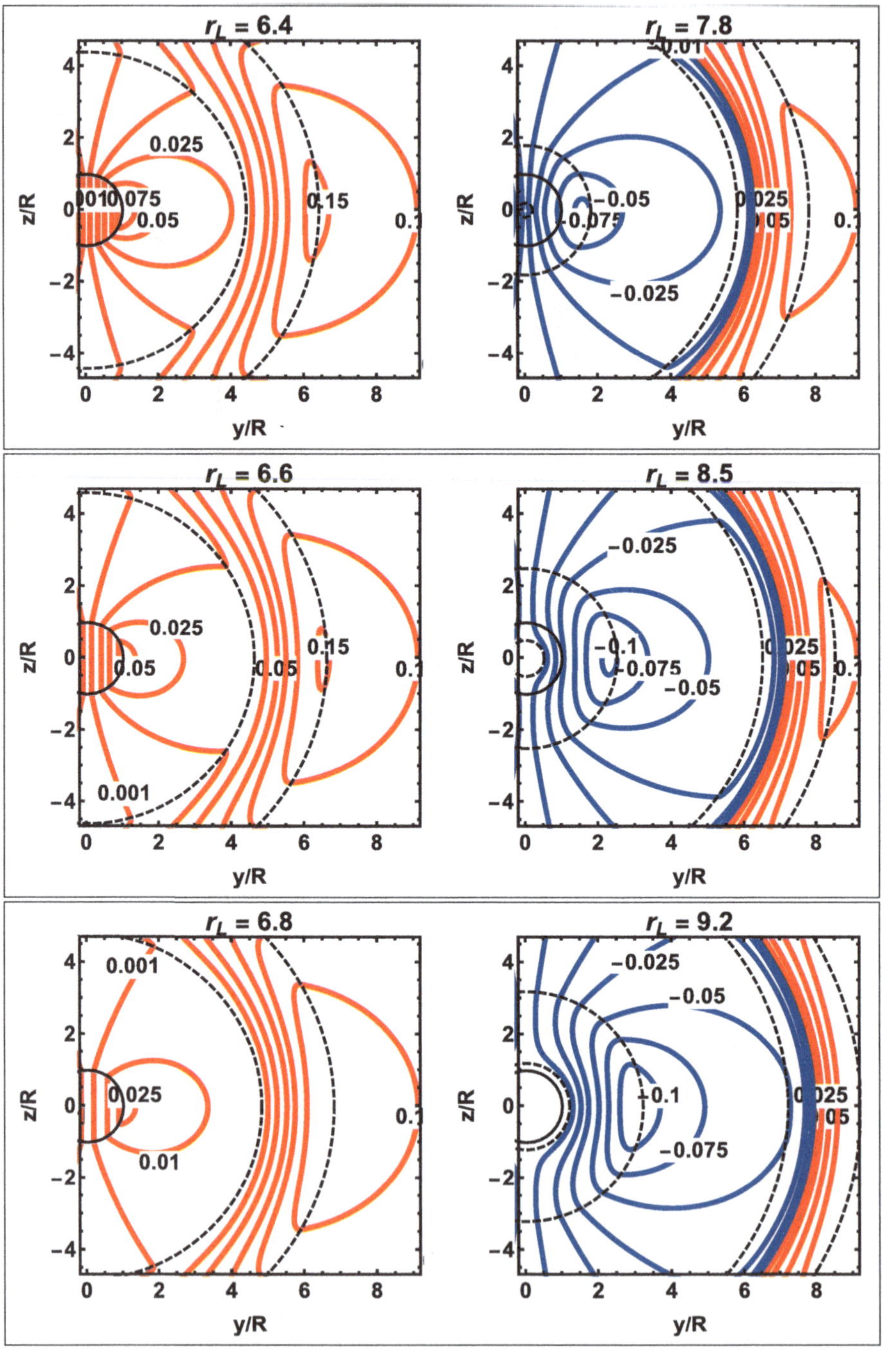

Fig. 6.5 Field lines for a linear off-ramp of width $\Delta r = 6$, showing the splitting and decay of the initial dipole field (red), and the growth of the outgoing bubble of blue flux. (Left) $r_L = 6.4$–6.8,

We have seen that the space-time maps are described by three equations. $f/r = \Phi/\Phi_0$ gives the red space-time tracks. $f - rf' = 0$ gives the black locus of the neutral points. And $f'' = 0$ combines with $f - rf' = 0$ to determine the coordinates of the black dot where the neutral points are born. Next, we use these equations to derive expressions for these coordinates, (r, r_L), and the flux fraction, Φ/Φ_0, of the track that passes through that point, in terms of the ramp width, Δr.

We begin by replacing the variables (r, r_L) by new variables (x, u), defined by $u = r_L - \Delta r - 1$ and $x = r - u$. Here, u is the midpoint of the trailing sheath (2o) and x is the radial distance from this midpoint, with positive values corresponding to positions ahead of the midpoint. With these definitions, $F_{2o}(r, r_L)$ from Eq. (5.9) becomes

$$F_{2o}(x, u) = 1 - \frac{(x+1)^2}{16\Delta r}\Big[3(x-1)^2 + 4(x-1)u - 4u\Big], \tag{6.6}$$

and the two conditions for the birth of the neutral point become $F_{2o} - (x + u)\,\partial F_{2o}/\partial x = 0$ and $\partial^2 F_{2o}/\partial x^2 = 0$. After some algebra, we obtain

$$u = (1 - 3x^2)/2x, \tag{6.7a}$$

$$16x^2\Delta r = (1+x)^3(3 - 13x + 18x^2 - 6x^3), \tag{6.7b}$$

where Eq. (6.7a) was derived from $\partial^2 F_{2o}/\partial x^2 = 0$, and Eq. (6.7b) was obtained by substituting the expression for u into $F_{2o} - (x + u)\,\partial F_{2o}/\partial x = 0$.

The numerical solution of these equations when $\Delta r = 3$ is $x = 0.197$ and $u = 2.237$, which give $r = x + u = 2.434$ and $r_L = 4 + u = 6.237$. When these coordinates are used to evaluate $F_{2o}(r, r_L)/r$, we obtain a flux fraction of 0.585. This contour has an asymptotic value of $r = 1/0.585 = 1.709$, which lies just above track #1 in Fig. 6.1, where one would expect a track with vertical slope.

We can also solve these equations analytically using a power series representation for x. First, we recognize in Eq. (6.7b) that $x \to 0$ as $\Delta r \to \infty$. In this case, the leading term in the series for x is obtained by setting $x = 0$ in the right side of Eq. (6.7b) giving $16x^2\Delta r = 3$. Consequently, $x \approx (1/4)\,(3/\Delta r)^{1/2}$ to first order in $(3/\Delta r)^{1/2}$. The corresponding value of $u = (1-3x^2)/2x \approx 1/2x = 2\,(3/\Delta r)^{-1/2}$. Next, we express both x and u as power series in $(3/\Delta r)^{1/2}$. By substituting these expressions into Eqs. (6.7a) and (6.7b), we can find the coefficients that make the resulting two series begin at arbitrarily large powers of $(3/\Delta r)^{1/2}$. Expanding to fourth order in $(3/\Delta r)^{1/2}$, but showing only the first three terms, we have

Fig. 6.5 (continued) showing the annihilation of flux at $r = 1$ and at the front of the leading sheath. (Right) $r_L = 7.8$–9.2, showing the emergence of blue flux toward the front of the trailing sheath and the evacuation of flux behind this sheath

$$x = \frac{1}{4}\left(\frac{3}{\Delta r}\right)^{1/2} - \frac{1}{24}\left(\frac{3}{\Delta r}\right) - \frac{1}{36}\left(\frac{3}{\Delta r}\right)^{3/2}, \tag{6.8a}$$

$$u = 2\left(\frac{3}{\Delta r}\right)^{-1/2} + \frac{1}{3} - \frac{7}{72}\left(\frac{3}{\Delta r}\right)^{1/2}, \tag{6.8b}$$

corresponding to

$$r = 2\left(\frac{3}{\Delta r}\right)^{-1/2} + \frac{1}{3} + \frac{11}{72}\left(\frac{3}{\Delta r}\right)^{1/2}, \tag{6.9a}$$

$$r_L = (\Delta r + 1) + 2\,(\frac{3}{\Delta r})^{-1/2} + \frac{1}{3} - \frac{7}{72}\left(\frac{3}{\Delta r}\right)^{1/2}. \tag{6.9b}$$

Substituting these series expansions for r and r_L into the flux function, $F_{2o}(r, r_L)/r$, and expanding the result to $O(3/\Delta r)^{3/2}$, we obtain

$$\Phi/\Phi_0 = \frac{1}{2}\left(\frac{3}{\Delta r}\right)^{1/2} + \frac{1}{12}\left(\frac{3}{\Delta r}\right) + \frac{1}{144}\left(\frac{3}{\Delta r}\right)^{3/2} \tag{6.10}$$

for the contour that passes through the point (r_L, r).

As anticipated, for large Δr, the variable x is small so that the neutral point is nearly centered in the trailing sheath where $r = r_L - \Delta r - 1$. For $\Delta r = 3$, these equations give $r \approx 2.49$ and $r_L \approx 6.24$, close to the exact solution, even though $3/\Delta r = 1$. When $\Delta r = 6$ as it was for the on-ramp in the left panel of Fig. 6.4, we obtain $r \approx 3.270$ and $r_L \approx 10.093$, within about 2% of the more exact numerical calculations ($r = 3.338$ and $r_L = 10.161$). Likewise, for $\Delta r = 6$, the series expansion for Φ/Φ_0 is 0.398, in close agreement with the value of 0.396 for the contour passing through this critical point in the left panel of Fig. 6.4. For an on-ramp of width $\Delta r = 10$, the approximations give $r = 4.068$, $r_L = 14.931$, and $\Phi/\Phi_0 = 0.300$, in good agreement with the values shown in Fig. 6.2.

Analogous neutral-point equations arise for the linear off-ramp with the result that

$$r = 2\left(\frac{3}{\Delta r}\right)^{-1/2} - \frac{1}{3} + \frac{11}{72}\left(\frac{3}{\Delta r}\right)^{1/2}, \tag{6.11a}$$

$$r_L = 1 + 2\left(\frac{3}{\Delta r}\right)^{-1/2} - \frac{1}{3} - \frac{7}{72}\left(\frac{3}{\Delta r}\right)^{1/2}, \tag{6.11b}$$

for the birthplace of the neutral points, and

$$\Phi/\Phi_0 = \frac{1}{2}\left(\frac{3}{\Delta r}\right)^{1/2} - \frac{1}{12}(\frac{3}{\Delta r}) + \frac{1}{144}\left(\frac{3}{\Delta r}\right)^{3/2} \tag{6.12}$$

for the field line passing through that point. When $\Delta r = 6$, these equations give $r = 2.603$, $r_L = 3.426$, and $\Phi/\Phi_0 = 0.314$, comparable to the values obtained with the more exact numerical solution ($r = 2.613$, $r_L = 3.428$, and $\Phi/\Phi_0 = 0.315$, as shown in the right panel of Fig. 6.4).

Here, we note that the on-ramp and off-ramp series for r differ only in the sign of the middle term. The same is true for the series for Φ/Φ_0 (and for r_L after taking account of the extra additive terms). Also, note that the leading terms of r and Φ/Φ_0 have reciprocal relations with $r \sim 2\sqrt{\Delta r/3}$ and $\Phi/\Phi_0 \sim (1/2)\sqrt{3/\Delta r}$.

This reciprocal relation reflects the fact that Φ_0/Φ is the radial position of the loop when it reaches its final location in the dipole where $F(r, r_L) = 1$. In this case, the difference $r - (\Phi_0/\Phi) \approx 2/3 + (1/8)(3/\Delta r)^{1/2}$ is the "falling distance" from the position, r, where the neutral points are born, to the final location in the dipole field. (For the off-ramp, Φ_0/Φ is the initial location of the loop, and the "falling distance" is $(\Phi_0/\Phi) - r \approx 2/3 - (1/8)(3/\Delta r)^{1/2}$).

Finally, we emphasize that the formation of the neutral points is only one aspect of the detachment process which also includes the weakening of the field, the separation of meridional field lines, and the collapse of closed loops prior to the formation of those neutral points. For the on-ramp, this occurs when the trailing sheath enters the field of view outside the sphere. The collapsing stops after the passage of the trailing boundary of the sheath. For the off-ramp, the collapsing occurs as the leading sheath moves outside the sphere and continues until $r_L = \Delta r + 1$ when the collapsing fields are annihilated by their rising counterparts inside the sphere.

Chapter 7
Non-linear Current Ramps

7.1 General Ideas

In the previous chapter, we described the magnetic field that occurs when an electric current is increased linearly with time or decreased linearly with time during a finite interval Δr. For this purpose, we regarded the linear on-ramp as a sequence of infinitesimal steps, each of whose individual contributions consists of an outgoing 'bubble' of closed loops followed by a correspondingly infinitesimal static dipole field, as described in Chap. 3. In the region of overlap, the bubbles interfered destructively, leaving a growing dipole field produced by the accumulation of the infinitesimal dipole contributions. But at the two ends of the wave train, the leading part of one bubble did not overlap with the trailing part of another bubble, so that a region of positive meridional flux was produced at the front of the wave train and a region of negative meridional flux was produced at the end. The result was a growing dipole field bounded by $2R$-wide sheaths of oppositely directed meridional field at each end. In this chapter, we extend the analysis to current ramps that are not linear, but are curved at one or both ends.

We begin by considering a non-linear current profile, $G(\xi)$, that varies monotonically from 0 to 1 for ξ in the range $(0, \Delta r)$. A simple example would be $G(\xi) = (\xi/\Delta r)^2$, which increases gradually at first and stops abruptly at the end of the interval. To obtain the f-values for such a profile, we return to Eqs. (5.6) and (5.7), and replace the infinitesimal steps of constant height, $dr'_L/\Delta r$, by infinitesimal steps of variable height, $G'(r_L - r'_L)dr'_L$. With this substitution, the integral $\int_{r_L-\Delta r}^{r_L} f_o(r, r'_L)\, dr'_L/\Delta r$ becomes $\int_{r_L-\Delta r}^{r_L} f_o(r, r'_L)\, G'(r_L - r'_L)\, dr'_L$, which equals $\int_0^{\Delta r} f_o(r, r_L - \xi)\, G'(\xi)\, d\xi$ when ξ is substituted for $r_L - r'_L$. We used the argument $r_L - r'_L$ in the function G' so that $r'_L = r_L$ would correspond to the start

Electronic Supplementary Material The online version of this chapter (https://doi.org/10.1007/978-3-030-40264-8_7) contains supplementary material, which is available to authorized users.

N. R. Sheeley, Jr., *Transient Magnetic Fields*,
https://doi.org/10.1007/978-3-030-40264-8_7

of the ramp at $\xi = 0$, and $r'_L = r_L - \Delta r$ would correspond to the end of the ramp at $\xi = \Delta r$.

In this way, Eqs. (5.6) and (5.7) can be systematically modified to give f-values for the non-linear ramps when the current is turned on. Expressions for off-ramps like $G(\xi) = 1 - (\xi/\Delta r)^2$ are then obtained by subtracting the on-ramp f-values from the f-values of the uniformly magnetized sphere that is left behind the on-ramp field (1 outside the sphere and r^3 inside). The evaluation of the resulting integrals is straightforward, but gives lengthy algebraic expressions similar to those in Eqs. (5.8) and (5.9). For most of these cases, we will not show those algebraic details here, but instead we will concentrate on the field-line plots and space-time maps generated from them. But first, we examine the continuity properties of the f-values and their derivatives for a general current-ramp profile, $G(\xi)$. We will find that F and $\partial F/\partial r$ are always continuous between the $1o/mo$ and $mo/2o$ boundaries and that the continuity of $\partial^2 F/\partial r^2$ depends on whether the current profile has zero slope at that end of the ramp.

First, consider the $1o/mo$ boundary for which $r = r_L - 2$. Using the substitution $dr'_L/\Delta r \to G'(r_L - r'_L)dr'_L$ in Eq. (5.7), and setting $r_L - r'_L = \xi$, we obtain

$$\begin{aligned} F_{1o}(r, r_L) &= \textstyle\int_0^{r_L - r} f_o(r, r_L - \xi)\, G'(\xi)\, d\xi, \\ F_{mo}(r, r_L) &= \textstyle\int_{r_L-(r+2)}^{r_L - r} f_o(r, r_L - \xi)\, G'(\xi)\, d\xi + \int_0^{r_L-(r+2)} G'(\xi)\, d\xi. \end{aligned} \tag{7.1}$$

Here, we see that when $r = r_L - 2$, the second integral in the expression for $F_{mo}(r, r_L)$ vanishes and the first integral in this expression becomes equal to the integral expression for $F_{1o}(r, r_L)$. Consequently, $F_{1o}(r, r_L) = F_{mo}(r, r_L)$ at the $1o/mo$ boundary, independent of the shape of the $G(\xi)$ profile. The same result is obtained at the other boundaries. In retrospect, this makes sense because F is proportional to the radial component of the field, B_r, whose continuity follows directly from $\nabla \cdot \mathbf{B} = 0$.

Next, we consider the first derivatives of these F-values with respect to r. Taking the partial derivatives of these two functions, we obtain

$$\begin{aligned} \partial F_{1o}(r, r_L)/\partial r &= \int_0^{r_L - r} \frac{\partial f_o(r, r_L - \xi)}{\partial r}\, G'(\xi)\, d\xi - f_o(r, r)\, G'(r_L - r), \\ \partial F_{mo}(r, r_L)/\partial r &= \int_{r_L-(r+2)}^{r_L - r} \frac{\partial f_o(r, r_L - \xi)}{\partial r}\, G'(\xi)\, d\xi - f_o(r, r)\, G'(r_L - r) \\ &\quad + f_o(r, r+2)\, G'(r_L - r - 2) - G'(r_L - r - 2). \end{aligned} \tag{7.2}$$

Referring to the definition of $f_o(r, r_L)$ in Eq. (2.14a), we see that $f_o(r, r) = 0$ and $f_o(r, r + 2) = 1$, in which case $\partial F_{1o}(r, r_L)/\partial r$ and $\partial F_{mo}(r, r_L)/\partial r$ become

$$\partial F_{1o}(r, r_L)/\partial r = \int_0^{r_L - r} \frac{\partial f_o(r, r_L - \xi)}{\partial r} G'(\xi)\, d\xi,$$
$$\partial F_{mo}(r, r_L)/\partial r = \int_{r_L-(r+2)}^{r_L - r} \frac{\partial f_o(r, r_L - \xi)}{\partial r} G'(\xi)\, d\xi. \tag{7.3}$$

When $r = r_L - 2$, these integrals are equal and therefore $\partial F/\partial r$ is continuous across the $1o/mo$ boundary independent of the shape of the profile, $G(\xi)$. The same result is obtained at the other boundaries. Now, because F and $\partial F/\partial r$ are both continuous across these boundaries, the quantity $F - r\, \partial F/\partial r$ is also continuous. Because $B_\theta \sim F - r\partial F/\partial r$ (as indicated in Eq. (3.3b)), it follows that B_θ is also continuous across these boundaries, independent of the shape of the current profile, $G(\xi)$.

However, $\partial^2 F(r, r_L)/\partial r^2$ and $\partial B_\theta/\partial r$ are not continuous across the $1o/mo$ boundary unless $G'(0) = 0$, and they are not continuous across the $2o/mo$ boundary unless $G'(\Delta r) = 0$. To show this, we return to Eq. (7.3) and take the derivatives as follows:

$$\partial^2 F_{1o}(r, r_L)/\partial r^2 = \int_0^{r_L - r} \frac{\partial^2 f_o(r, r_L - \xi)}{\partial r^2} G'(\xi)\, d\xi$$
$$-\partial f_o(r, r_L - \xi)/\partial r|_{\xi = r_L - r}\, G'(r_L - r),$$
$$\partial F_{mo}(r, r_L)/\partial r = \int_{r_L-(r+2)}^{r_L - r} \frac{\partial^2 f_o(r, r_L - \xi)}{\partial r^2} G'(\xi)\, d\xi$$
$$-\partial f_o(r, r_L - \xi)/\partial r|_{\xi = r_L - r}\, G'(r_L - r)$$
$$+\partial f_o(r, r_L - \xi)/\partial r|_{\xi = r_L - r - 2}\, G'(r_L - r - 2). \tag{7.4}$$

Again, referring to the definition of $f_o(r, r_L)$ in Eq. (2.11a), we can easily show that the quantity $\partial f_o(r, r_L - \xi)/\partial r|_{\xi = r_L - r}$ is $-3r/2$ and also that $\partial f_o(r, r_L - \xi)/\partial r|_{\xi = r_L - r - 2}$ is $+3r/2$. In this case, Eq. (7.1) reduces to

$$\partial^2 F_{1o}(r, r_L)/\partial r^2 = \int_0^{r_L - r} \frac{\partial^2 f_o(r, r_L - \xi)}{\partial r^2} G'(\xi)\, d\xi + \frac{3}{2} r\, G'(r_L - r),$$
$$\partial^2 F_{mo}(r, r_L)/\partial r^2 = \int_{r_L-(r+2)}^{r_L - r} \frac{\partial^2 f_o(r, r_L - \xi)}{\partial r^2} G'(\xi)\, d\xi + \frac{3}{2} r\, G'(r_L - r)$$
$$+\frac{3}{2} r\, G'(r_L - r - 2). \tag{7.5}$$

When $r = r_L - 2$, the difference between these two quantities is $(3r/2)\, G'(0)$, which vanishes if $G'(0) = 0$. Consequently, $G'(0) = 0$ is the condition that $\partial^2 F/\partial r^2$ be continuous across the $1o/mo$ boundary. A similar argument shows that $G'(\Delta r) = 0$ is the condition that $\partial^2 F/\partial r^2$ be continuous across the $mo/2o$ boundary. Because $\partial^2 F/\partial r^2 = -(1/r)\, \partial(F - r\partial F/\partial r)/\partial r$, it follows that these are also the conditions that $\partial B_\theta/\partial r$ be continuous across the $1o/mo$ and $mo/2o$ boundaries.

7.2 Quadratic/Linear (QL) and Linear/Quadratic (LQ) Profiles

Figure 7.1 shows some of these non-linear current profiles. We describe these profiles in terms of a function, $G(\xi)$, whose independent variable, ξ, represents time in the range $(0, \Delta r)$. In the upper left, the function $G(\xi) = (\xi/\Delta r)^2$ begins quadratically and ends linearly, in the sense that the first non-zero (and non-constant) terms in the expansion of $G(\xi)$ about the starting and ending points are quadratic and linear, respectively. We shall refer to this profile as a quadratic/linear (QL) on-ramp. In the upper right, the ramp also begins quadratically, but this ramp curves downward toward its linear end, and therefore is a QL off-ramp. The bottom panels show the corresponding LQ on-ramp and off- ramp. Note that the on-ramps run from 0 to 1 and the off-ramps run from 1 to 0.

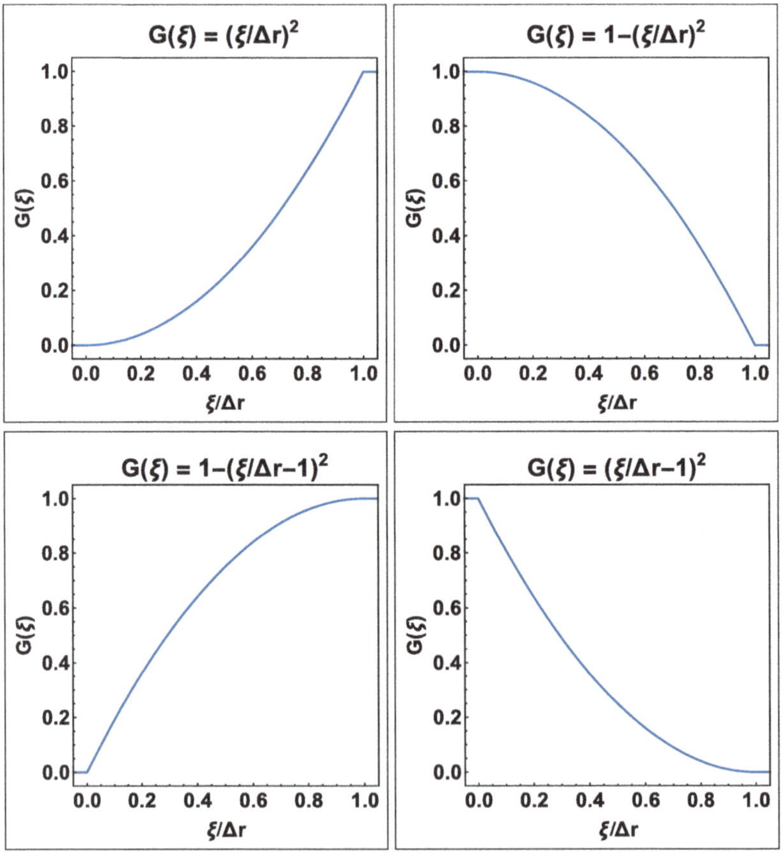

Fig. 7.1 Non-linear current profiles for on-ramps (left) and off-ramps (right). (Top) Quadratic start and linear finish (QL), (bottom) linear start and quadratic finish (LQ)

Now, we can predict that the fields for the QL profiles in the upper panels of Fig. 7.1 will have continuous derivatives, $\partial B_\theta/\partial r$, across the $1o/mo$ boundary, but not across the $mo/2o$ boundary. Likewise, the fields for the LQ profiles in the lower panels will have continuous derivatives $\partial B_\theta/\partial r$ across the $mo/2o$ boundary, but not across the $1o/mo$ boundary. Thus, despite the use of dashed lines to show these boundaries in space-time maps and field line plots, we recognize that B_θ and its derivative, $\partial B_\theta/\partial r$, are continuous across the leading $(1o/mo)$ boundaries of QL maps and across the trailing $(mo/2o)$ boundaries of LQ maps.

Figure 7.2 shows the space-time maps corresponding to the non-linear current ramps in Fig. 7.1. In each case, we use a ramp length of $\Delta r = 6$ to facilitate a comparison with the space-time maps for the linear ramps in Fig. 6.4. As one might expect, the space-time maps for these non-linear ramps are similar to the space-time maps for the corresponding linear ramps in the regions where the profiles are approximately linear (i.e., at the trailing ends of the QL ramps and at the leading ends of the LQ ramps).

In the upper left panel of Fig. 7.2, the space-time map for the QL on-ramp shows neutral points evolving in the same way that they did for the linear on-ramps in Figs. 6.2 and 6.4. These neutral points are born approximately midway in the trailing sheath and migrate toward the front and back of the sheath. The behavior is much different for the LQ on-ramp in the lower left panel of Fig. 7.2. Here, the neutral points are born at the back end of the leading sheath. The O-type sink point remains in the leading sheath and moves outward very close to the back boundary of that sheath. However, the X-type source point drifts slowly across the middle region before crossing the $mo/2o$ boundary and accelerating rapidly outward toward the back of the trailing sheath.

For the off-ramps in the right panels, the neutral points are born in the leading sheaths. In both the QL-off and LQ-off maps, the O-type sink point moves asymptotically outward along the front of the leading sheath. For the LQ off-ramp, the X-type neutral point remains in the leading sheath, initially providing a source of newly pinched field lines. However, when all of these field lines are pinched off, the neutral point changes to an O-type point where blue loops of reversed polarity emerge along the back of the leading sheath. In contrast, for the QL off-ramp, the X-type source point moves inward across the middle region and then changes to an O-type source of reversed-polarity flux along the front of the trailing region.

We can summarize this behavior as follows. First, we recognize that the Pinchoff occurs in the negative part of the outgoing field, which is the trailing part of the on-ramp field, but the leading part of the off-ramp field. Consequently, for the QL on-ramp and the LQ off-ramp, the Pinchoff is determined by the linear part of the current profile and occurs in narrow sheaths as it did for the linear ramps. However, for the LQ on-ramp and the QL off-ramp, the Pinchoff is determined by the quadratic part of the current profile, and therefore spreads broadly across the middle region before ending up in the trailing sheath.

Figure 7.2 showed that for the LQ on-ramp, the neutral point formed at the $1o/mo$ boundary where $r = r_L - 2$ rather than in a sheath where $f''(r, r_L) = 0$, as happened for the other cases. Consequently, we decided to show the fields of the LQ

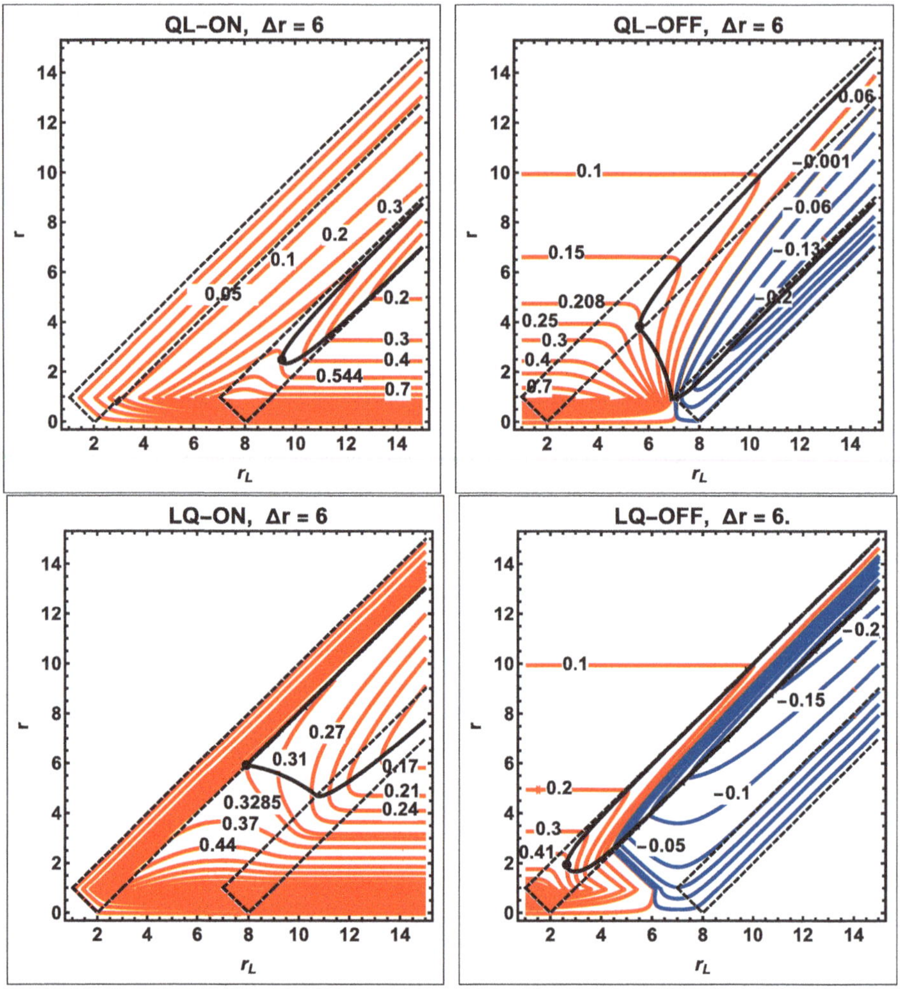

Fig. 7.2 Contours of constant flux for the non-linear current profiles given in the previous figure for a width $\Delta r = 6$. Dotted boxes indicate the boundaries of the leading and trailing sheaths. Solid black curves show the *loci* of neutral points where the curves have vertical slopes. The black dot separates each curve into a lower segment where flux emerges and an upper segment where flux is annihilated. Red and blue colors indicate field lines whose circulation is clockwise and counterclockwise, respectively (when $y > 0$)

on-ramp in more detail. Figure 7.3 shows these fields during $r_L = 8.92$–13.54 when the leading and trailing sheaths crossed the field of view. In the left panels, the field line marked 0.31 pinches off inside the *mo* region. In the lower left panel of Fig. 7.2, this location occurs where the horizontal segment of the black neutral-point curve intersects the red 0.31 contour.

Returning to Fig. 7.3, we can see that the 0.31 field line was separated into a collapsing loop and the back end of an outgoing closed loop. As the trailing sheath

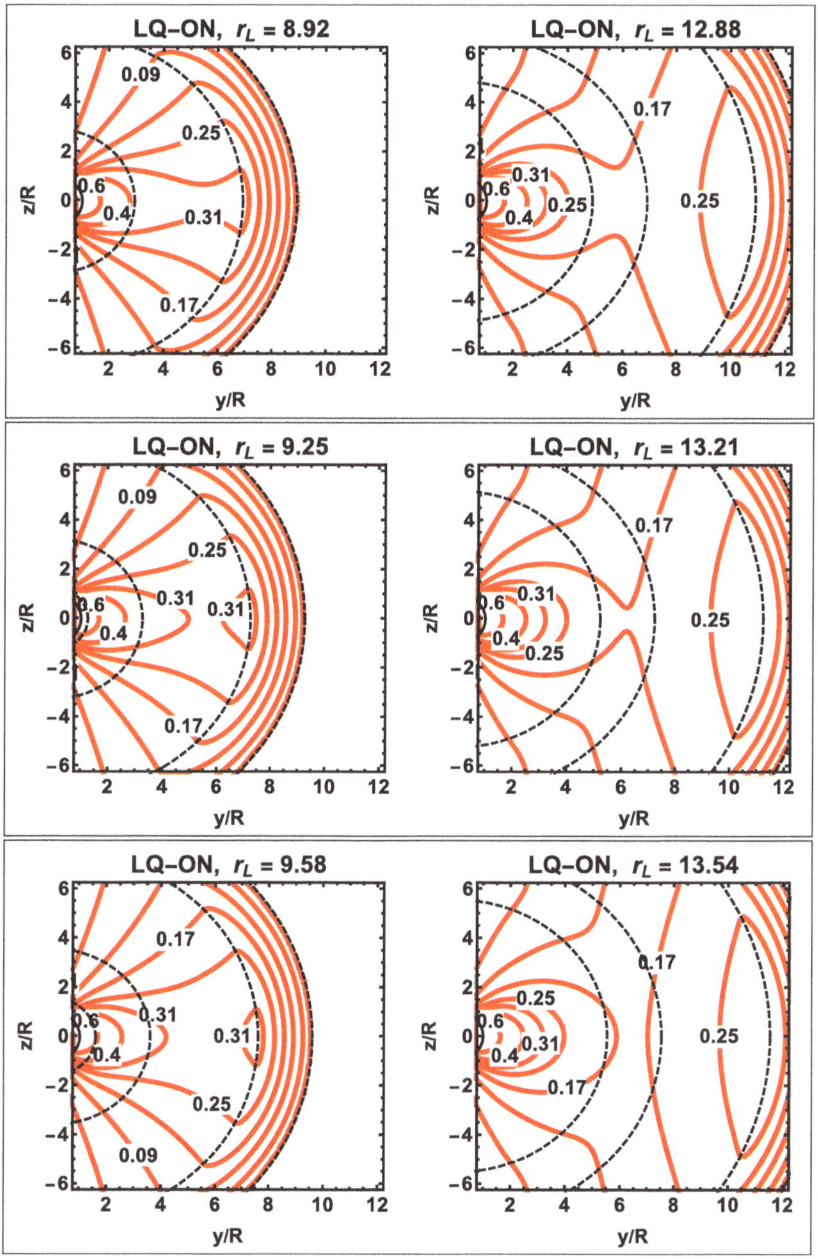

Fig. 7.3 Field lines for an LQ on-ramp of width $\Delta r = 6$, showing the Pinchoff of the $\Phi = 0.31$ contour in the middle region (left), and the Pinchoff of the $\Phi = 0.17$ contour in the trailing sheath (right). Each Pinchoff separates a stretched field line into an ingoing loop that becomes part of the dipole and an outgoing loop that annihilates its counterpart in the leading sheath, producing an asymmetric circulation of flux

passed by, the downward motion ended and the loop joined the static loops of the dipole field, as seen in the right panels of Fig. 7.3. The outgoing loop rapidly decreased in size and disappeared at the back end of the leading sheath.

In the right panels of Fig. 7.3, a loop marked 0.17 pinched off near the center of the trailing sheath at $r_L = 13.21$. In the lower left panel of Fig. 7.2, this location occurs where the outgoing segment of the neutral-point curve intersects the 0.17 contour. Returning to Fig. 7.3, the last panel at $r_L = 13.54$ provides a hint of the final outgoing structure, which seems to be an asymmetric bubble with meridional flux concentrated in the leading sheath and its return flux spread widely across the middle and trailing sheath. As we shall see in the next figure, such asymmetric bubbles are characteristic of these asymmetric current ramps.

Figure 7.4 shows field lines for the four asymmetric current ramps when $r_L = 34.5$. In each case, the ramp width is $\Delta r = 12$. Also, the leading and trailing sheaths are located at radial distances of 32.5–34.5 and 20.5–22.5, respectively, as indicated by the dashed circles. A glance at Fig. 7.4 reveals that the trailing flux is more concentrated than the leading flux for the QL fields in the top panels, and that the reverse is true for the LQ fields in the bottom panels. Thus, the more concentrated flux accompanies the linear part of the profile and the less concentrated flux accompanies the quadratic part of the profile, regardless of which occurs first.

We can gain further insight by comparing the field lines in Fig. 7.4 with the corresponding space-time maps in Fig. 7.2 (even though the space-time maps were calculated for $\Delta r = 6$). For the field of the QL on-ramp in Fig. 7.4, the concentration of flux in the trailing sheath is bounded by the closely spaced neutral points at the back of the sheath (where flux is created) and at the front of the sheath (where flux is destroyed). The same is true for the LQ off-ramp. In each case, the return flux is spread over the middle region (mo) and the more distant sheath.

For the field of the LQ on-ramp, the neutral points are more widely separated, occurring at the back ends of the two sheaths. Again, flux is created at the back end of the trailing sheath and destroyed at the back end of the leading sheath. Consequently, the meridional field is relatively weak over this broad region, but stronger in the leading sheath. Likewise, the neutral points are widely separated for the field of the QL off-ramp. In this case, they occur at the front ends of the sheaths, with flux being created in the trailing sheath and destroyed at the leading sheath. So the field is weak in the broad region between the neutral points and strong in the trailing sheath.

In summary, the circulating meridional flux is spread between the two neutral points, being concentrated when the neutral points are close together in the same narrow sheath (QL-on and LQ-off) and diluted when the neutral points are far apart, either at the back edges of the sheaths (QL-off) or the front edges of the sheaths (LQ-on).

Next, in our analysis of the fields of the QL and LQ profiles, we list the values of F_{mo} obtained for the middle region using the generalized expressions corresponding to Eq. (5.7) with $G'(r_L - r'_L)dr'_L$ substituted for $dr'_L/\Delta r$. After some algebra, we obtain

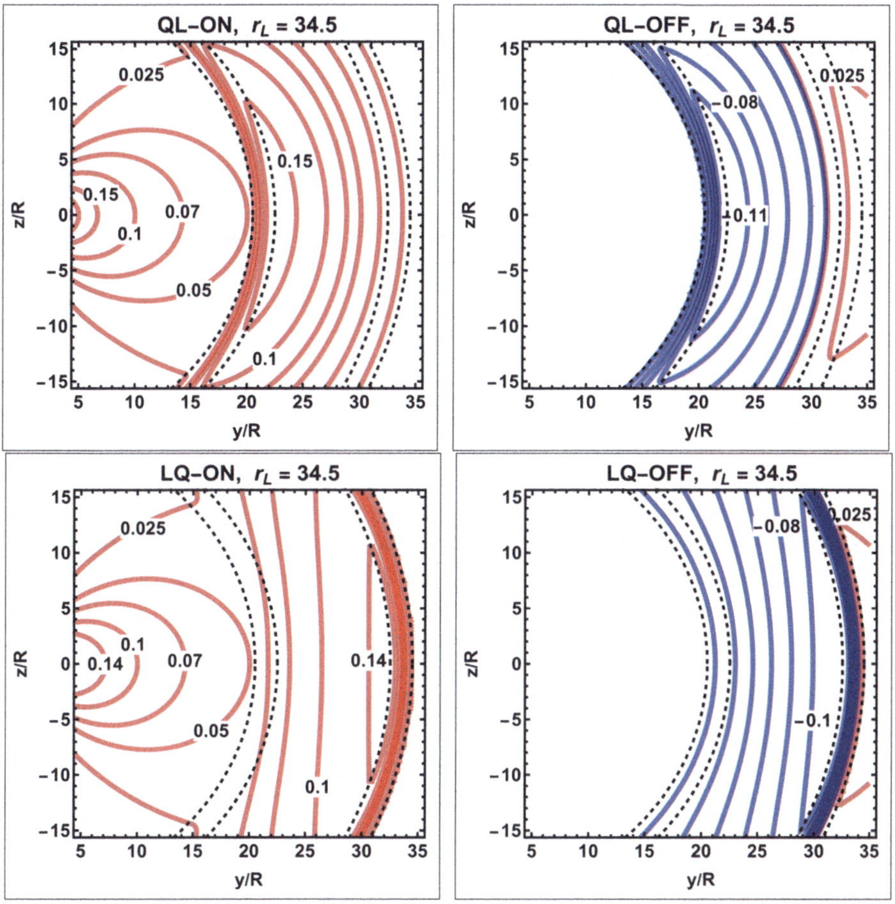

Fig. 7.4 Field lines for the non-linear current profiles with a width $\Delta r = 12$, showing the asymmetric shapes of the bubbles when $r_L = 34.5$. (Top) Field lines for the QL on-ramp (left) and the QL off-ramp (right), (bottom) corresponding field lines for the LQ on- and LQ off-ramps. Dotted circles indicate the leading and trailing sheaths. Red and blue colors indicate field lines whose circulation is clockwise and counterclockwise, respectively (when $y > 0$)

$$F_{mo}(r, r_L) = \begin{cases} \dfrac{1}{5(\Delta r)^2} + \dfrac{(r_L-1)^2}{(\Delta r)^2} - \dfrac{r^2}{(\Delta r)^2}, & \text{QL-on} \\ 1 - \left[\dfrac{1}{5(\Delta r)^2} + \dfrac{(r_L-1)^2}{(\Delta r)^2} - \dfrac{r^2}{(\Delta r)^2}\right], & \text{QL-off} \\ \dfrac{1}{5(\Delta r)^2} + \dfrac{(r_L-\Delta r-1)^2}{(\Delta r)^2} - \dfrac{r^2}{(\Delta r)^2}, & \text{LQ-off} \\ 1 - \left[\dfrac{1}{5(\Delta r)^2} + \dfrac{(r_L-\Delta r-1)^2}{(\Delta r)^2} - \dfrac{r^2}{(\Delta r)^2}\right]. & \text{LQ-on} \end{cases} \tag{7.6}$$

From these equations, we can easily derive the values of $F_{mo} - rF'_{mo}$, and then use them to interpret the formation of the neutral points and associated Pinchoffs of the field shown in the panels of Fig. 7.2. After more algebra, we obtain

$$F_{mo}(r, r_L) - rF'_{mo}(r, r_L) = \begin{cases} \frac{1}{5(\Delta r)^2} + \frac{(r_L-1)^2}{(\Delta r)^2} + \frac{r^2}{(\Delta r)^2}, & \text{QL-on} \\ 1 - \left[\frac{1}{5(\Delta r)^2} + \frac{(r_L-1)^2}{(\Delta r)^2} + \frac{r^2}{(\Delta r)^2}\right], & \text{QL-off} \\ \frac{1}{5(\Delta r)^2} + \frac{(r_L-\Delta r-1)^2}{(\Delta r)^2} + \frac{r^2}{(\Delta r)^2}, & \text{LQ-off} \\ 1 - \left[\frac{1}{5(\Delta r)^2} + \frac{(r_L-\Delta r-1)^2}{(\Delta r)^2} + \frac{r^2}{(\Delta r)^2}\right], & \text{LQ-on} \end{cases} \tag{7.7}$$

where $F'_{mo}(r, r_L) = \partial F_{mo}(r, r_L)/\partial r$ as usual.

Note that the values of $F_{mo} - rF'_{mo}$ differ from the values of F_{mo} only by the sign of the r^2 term. This difference causes the values of $F_{mo} - rF'_{mo}$ to be positive for the QL-on and LQ-off cases and to be 1 minus those positive quantities for the other two cases. These values are analogous to the values of $(r_L - 1)/\Delta r$ and $1 - (r_L - 1)/\Delta r$ that we obtained for the linear on-ramp and off-ramp, respectively, as described above. Consequently, for the QL on-ramp and the LQ off-ramp, the neutral points cannot form in the middle region where $F_{mo} - rF'_{mo} > 0$, but instead must form in the trailing and leading sheaths where $F - rF'$ can vanish. On the other hand, for the QL off-ramp and the LQ on-ramp, the neutral points can move across the middle region because $F_{mo} - rF'_{mo}$ is of the form "1 minus a positive number" which can vanish there, giving the neutral-point curves shown in the lower left and upper right panels of Fig. 7.2.

In Chap. 6, we used a power series technique to deduce the space-time coordinates and field-line fluxes where the neutral points were born. Now, we will repeat this procedure for the QL and LQ current ramps of this section. For the QL on-ramp, we obtain

$$r = 2\left(\frac{6}{\Delta r}\right)^{-1/2} + \frac{1}{3} + \frac{19}{72}\left(\frac{6}{\Delta r}\right)^{1/2}, \tag{7.8a}$$

$$r_L = 2\left(\frac{6}{\Delta r}\right)^{-1/2} + \frac{1}{3} + \frac{1}{72}\left(\frac{6}{\Delta r}\right)^{1/2} + \Delta r + 1, \tag{7.8b}$$

$$\Phi/\Phi_0 = \frac{1}{2}\left(\frac{6}{\Delta r}\right)^{1/2} + \frac{1}{12}\left(\frac{6}{\Delta r}\right) - \frac{1}{48}\left(\frac{6}{\Delta r}\right)^{3/2}, \tag{7.8c}$$

and for the LQ off-ramp, we obtain the very similar terms:

$$r = 2\left(\frac{6}{\Delta r}\right)^{-1/2} - \frac{1}{3} + \frac{19}{72}\left(\frac{6}{\Delta r}\right)^{1/2}, \tag{7.9a}$$

$$r_L = 2\left(\frac{6}{\Delta r}\right)^{-1/2} - \frac{1}{3} + \frac{1}{72}\left(\frac{6}{\Delta r}\right)^{1/2} + 1, \tag{7.9b}$$

$$\Phi/\Phi_0 = \frac{1}{2}\left(\frac{6}{\Delta r}\right)^{1/2} - \frac{1}{12}\left(\frac{6}{\Delta r}\right) - \frac{1}{48}\left(\frac{6}{\Delta r}\right)^{3/2}. \tag{7.9c}$$

These expressions are quite different from what we obtain for the QL off-ramp and the LQ on-ramp. For the QL off-ramp, we obtain

$$r = \left(\frac{\sqrt{2}}{\Delta r}\right)^{-1} - \frac{1}{2} + \frac{13}{40}\left(\frac{\sqrt{2}}{\Delta r}\right) + \frac{14}{27}\left(\frac{\sqrt{2}}{\Delta r}\right)^{2}, \tag{7.10a}$$

$$r_L = \left(\frac{\sqrt{2}}{\Delta r}\right)^{-1} - \frac{1}{2} - \frac{41}{120}\left(\frac{\sqrt{2}}{\Delta r}\right) - \frac{1}{27}\left(\frac{\sqrt{2}}{\Delta r}\right)^{2} + 2, \tag{7.10b}$$

$$\Phi/\Phi_0 = \frac{\sqrt{2}}{\Delta r} - \frac{1}{2}\left(\frac{\sqrt{2}}{\Delta r}\right)^{2} - \frac{1}{120}\left(\frac{\sqrt{2}}{\Delta r}\right)^{3}, \tag{7.10c}$$

and for the LQ on-ramp, we obtain

$$r = \Delta r - \frac{3}{5}\left(\frac{1}{\Delta r}\right) + \frac{3}{5}\left(\frac{1}{\Delta r}\right)^{2} - \frac{24}{25}\left(\frac{1}{\Delta r}\right)^{3}, \tag{7.11a}$$

$$r_L = \Delta r - \frac{3}{5}\left(\frac{1}{\Delta r}\right) + \frac{3}{5}\left(\frac{1}{\Delta r}\right)^{2} - \frac{24}{25}\left(\frac{1}{\Delta r}\right)^{3} + 2, \tag{7.11b}$$

$$\Phi/\Phi_0 = \frac{2}{\Delta r} - \frac{6}{5}\left(\frac{1}{\Delta r}\right)^{3}. \tag{7.11c}$$

From these four sets of equations, we can see that the results occur in pairs. For the QL on-ramp and the LQ off-ramp, the neutral points are born at the linear end of the current profile, and the series expressions for their coordinates and fluxes have the same functional form as we obtained for the linear ramps in Chap. 6. Also, the "falling distances", $(2/3) \pm (1/8)(6/\Delta r)^{1/2}$, are essentially the same as we obtained for the linear ramps. The principal difference is that Eqs. (7.8) and (7.9) use the argument $6/\Delta r$, whereas Eqs. (6.9)–(6.10) and (6.11)–(6.12) use the argument $3/\Delta r$. This difference reflects the lower slope of the linear ramp, which is $1/\Delta r$, compared to $2/\Delta r$ at the linear end of the quadratic ramps. In general, for a monotonic ramp that is linear at one end and curved at the other end, the argument would be three times the slope at the linear end (i.e., $3n/\Delta r$ for a profile of the form $G(\xi) = (\xi/\Delta r)^n$).

However, looking at Eqs. (7.10) and (7.11), we can see that the situation is different for the QL off-ramp and the LQ on-ramp. The neutral points are born at the curved ends of these ramps, and the power series expansions depend on $1/\Delta r$ rather than on $(1/\Delta r)^{1/2}$, as they did for neutral points born at the linear ends of the ramps. The series expansions for the QL off-ramp and LQ on-ramp have two additional distinctions. First, the QL off-ramp has an extra factor of $\sqrt{2}$ in its argument, and its "falling distance" is approximately $1 - (1/15)(\sqrt{2}/\Delta r)$. Second, for the LQ on-ramp, the leading term in the series expansion for the flux factor, Φ/Φ_0, is not simply the reciprocal of the leading term in the series for r, as it is for all of the other cases. Instead, the leading term of Φ/Φ_0 is twice the reciprocal of the leading term in the series for r, and the "falling distance" is $(1/2)\Delta r - (9/10)(1/\Delta r)$. So the LQ on-ramp is unique is those respects.

But as a general rule, we can say that for ramps of moderate length, Δr, the coordinates vary roughly as $\sqrt{\Delta r}$ for neutral points that are born at the linear ends of the ramps and as Δr for neutral points that are born in the curved portion of the ramps. Likewise, the corresponding flux fractions are given approximately by the reciprocal of those terms, being $\sim 1/\sqrt{\Delta r}$ at the linear ends and $\sim 1/\Delta r$ at the curved ends.

As a final contribution to this subsection, we consider a more intuitive approach. Recall that in Fig. 5.4, we used a series of uniformly spaced circular fields to illustrate how bubbles from the infinitesimal steps of a linear current ramp combine to produce leading and trailing sheaths of oppositely directed meridional field. It is interesting to see how that analogy extends to our non-linear ramps. For this purpose, we refer to Fig. 7.5, which shows the result of combining wave trains

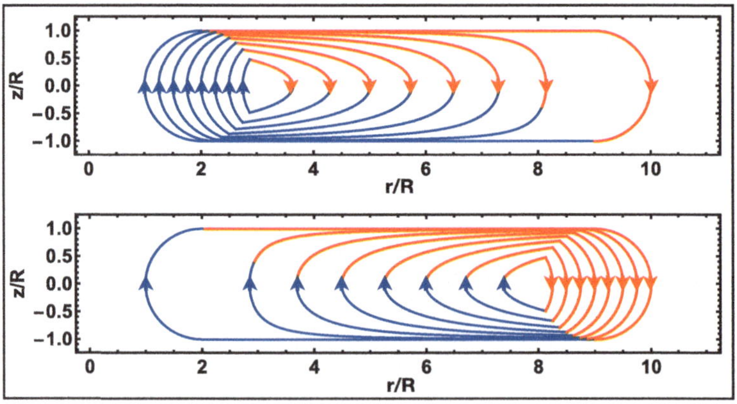

Fig. 7.5 Field lines produced by a series of circular loops moving to the right with $\Delta r = 7$ and $r_L = 10$, as in Fig. 5.4. Here, the original loop strengths are not constant, corresponding to $G'(\xi) = 1/\Delta r$, but vary linearly across the wave train, corresponding to $G'(\xi) = (2/\Delta r)(\xi/\Delta r)$ for the QL profile and $G'(\xi) = (2/\Delta r)(1 - \xi/\Delta r)$ for the LQ profile shown in the left panels of Fig. 7.1. Whereas the field lines in Fig. 5.4 have sharp boundaries at both ends of the wave train, the field lines in this figure lose their boundaries at the Q end where the fields spread continuously into the middle region

for circular fields with the QL-on and LQ-on profiles in the upper and lower panels, respectively. Unlike the strengths of the original loops, which were constant corresponding to $G'(\xi) = 1/\Delta r$, the strengths of these loops vary linearly across the wave train corresponding to $G'(\xi) = (2/\Delta r)(\xi/\Delta r)$ for the QL profile in the upper panel and $G'(\xi) = (2/\Delta r)(1 - \xi/\Delta r)$ for the LQ profile in the lower panel. Whereas the field lines in Fig. 5.4 had sharp boundaries at both ends of the wave train, the field lines here lose their boundaries at the quadratic (Q) end where the fields spread continuously into the middle region.

These idealized field-line patterns are useful for interpreting the space-time maps in Fig. 7.2. For example, consider the field of the idealized QL on-ramp in the upper panel of Fig. 7.5. Because it is the field of an on-ramp, the Pinchoff occurs at the end of the wave train where the profile is linear, the field is strong, and the slopes of the field lines are discontinuous at the boundaries of the sheath. It is easy to imagine that the strong negative field of this trailing sheath weakens the positive field created at the front and middle parts of the disturbance. In this case, a pair of neutral points would emerge at a point within the sheath and rapidly separate as they move toward the front and back of the sheath.

The polarities are the same for the field of the idealized LQ on-ramp in the lower panel, except that the leading fields are concentrated and the trailing fields are spread out. Consequently, the Pinchoff would occur in the trailing part of the wave train where the profile is quadratic, the field is weaker, and the slopes of the field lines are continuous at the boundaries of the trailing sheath. Thus, we would expect the neutral points to form close to the $1o/mo$ boundary where the negative field has its greatest strength. One neutral point might remain close to this boundary, while the other neutral point would move through the middle region and across the "continuous" $2o/mo$ boundary to approach the end of the wave train.

To derive these fields, we separate the regions into the left and right circular areas and the central region between them. Consider the fields in the upper panel. In the left and right circular areas, the fields are just the tangential fields of the individual circles and the field lines are portions of those circles that lie inside the first and last circle. The fields in the central region are obtained from the vector sum of the fields of two intersecting circles of differing field strengths. In the (r, z) rectangular coordinate system, the result is

$$\begin{aligned} B_r(r, z) &= -(4/\Delta r)[r - (\Delta r + 2)]\,(z/\Delta r) \\ B_z(r, z) &= -(4/\Delta r)\left(1 - z^2\right)/\Delta r. \end{aligned} \tag{7.12}$$

Setting $dz/dr = B_z(r, z)/B_r(r, z)$, and solving the resulting differential equation, we obtain

$$z^2 = 1 - k^2/[r - (\Delta r + 2)]^2, \tag{7.13}$$

where k is a constant of integration whose value can be obtained by matching a field line in this middle region to a corresponding circular arc in the left area. If the circle is shifted from the left end by the amount s, then $k(s) = (s/2)(\Delta r - s/2)$. Moreover, if $k(s) > 1$, then that middle field line will not reach the right circular area, but instead it will cross the r-axis at $r = \Delta r + 2 - k(s)$. In this example for which $\Delta r = 7$ and $s = 0.00, 0.25, 0.50, \ldots, 1.75$, only 2 of the 8 blue circles on the left side of the upper panel are linked to red circular arcs on the right side. The rest of the blue circles are closed by shorter loops in the middle area. A similar argument leads to the "reversed" pattern of field lines in the lower panel of Fig. 7.5.

7.3 Exponential Profiles

Next, we consider current profiles that rise or fall exponentially to their final values, as shown in Fig. 7.6. These profiles are particularly interesting for two reasons. First, they begin linearly and end quadratically and therefore resemble the LQ profiles in the lower panels of Fig. 7.1. Consequently, these exponential profiles might be expected to have space-time maps and field lines similar to those of the LQ profiles. Second, these profiles are what we would expect if a voltage source were turned on (or off) suddenly, causing the current to increase (or decrease) gradually with a time constant $1/\alpha$. Thus, for these relatively interesting examples, we shall provide the complete f-values both in the leading sheaths and after those sheaths, despite the relatively complicated mathematical appearance of those f-values.

Again, we obtain the f-values by replacing $dr'_L/\Delta r$ by $G'(r_L - r'_L)\, dr'_L$ in Eqs. (5.6) and (5.7). For the on-ramp shown in the left panel of Fig. 7.6, $G'(\xi) = \alpha e^{-\alpha\xi}$, where $1/\alpha$ is the decay time of the exponential. In this case, the f-values inside the sphere are

$$F_{1i}(r, r_L) = f_i(r, r_L) - (3/4\alpha)\left(r^2 - r_L^2\right) + \left(3/2\alpha^3\right)\left[(1+\alpha)(1-\alpha r_L)\right.$$

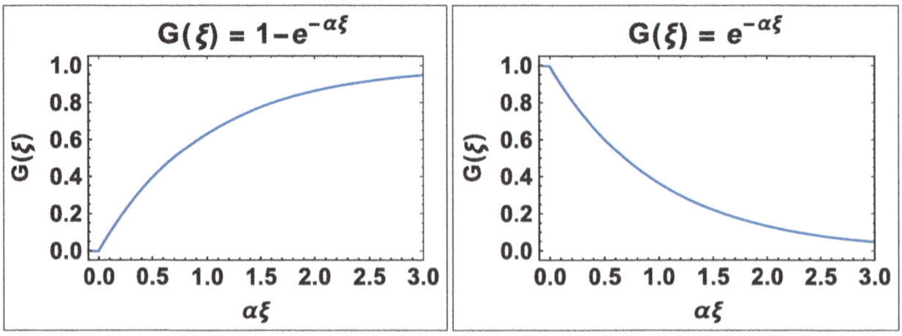

Fig. 7.6 Current profiles with a linear start and an exponential end. The on-ramp is on the left and the off-ramp is on the right

$$-(1-\alpha)(1+\alpha r)e^{-\alpha(r_L+r-2)}\Big],$$

$$F_{mi}(r, r_L) = r^3 \left[1 \ - \ (1-\alpha)\, h(\alpha r)\, e^{-\alpha(r_L-2)}\right], \tag{7.14}$$

and those outside the sphere are

$$F_{1o}(r, r_L) = f_o(r, r_L) - (3/4\alpha)\left(r^2 - r_L^2\right) + \left(3/2\alpha^3\right)(1+\alpha)\Big[(1-\alpha r_L)$$

$$-(1-\alpha r)e^{-\alpha(r_L-r)}\Big],$$

$$F_{mo}(r, r_L) = 1 \ - \ (1-\alpha r)\, h(\alpha)\, e^{-\alpha(r_L-r-1)}. \tag{7.15}$$

Here, f_o and f_i are the usual f-values for a sudden turn-on as given in Eqs. (2.11a) and (2.11b), respectively. The symbol, h, is an auxiliary function defined by

$$h(\xi) \ = \ \left(3/2\xi^3\right) \ \left[(1+\xi)\, e^{-\xi} \ - \ (1-\xi)\, e^{\xi}\right], \tag{7.16}$$

and has the property that $h(\xi) \approx 1 + \xi^2/10 + O(\xi^4)$ for small ξ. We do not include f-values for the $2o$ region because the f-values in the middle region (mo) continue indefinitely due to the exponential nature of the current profile.

As before, we obtain the f-values for the off-ramp by subtracting these on-ramp values from those of a uniformly magnetized sphere (i.e., r^3 inside the sphere and 1 outside the sphere). Inside the sphere, the values are

$$F_{1i}(r, r_L) = r^3 - f_i(r, r_L) - \left(3/2\alpha^3\right)\Big\{(1+\alpha)(1-\alpha r_L)$$

$$-(1-\alpha)(1+\alpha r)e^{-\alpha(r_L+r-2)}\Big\} + \ (3/4\alpha)\left(r^2 - r_L^2\right),$$

$$F_{mi}(r, r_L) = r^3\, (1-\alpha)\, h(\alpha r)\, e^{-\alpha(r_L-2)}, \tag{7.17}$$

and outside the sphere the values are

$$F_{1o}(r, r_L) = 1 - f_o(r, r_L) + (3/4\alpha)\left(r^2 - r_L^2\right) - \left(3/2\alpha^3\right)(1+\alpha)$$

$$\times \left\{(1-\alpha r_L) - (1-\alpha r)e^{-\alpha(r_L-r)}\right\},$$

$$F_{mo}(r, r_L) = (1-\alpha r)\, h(\alpha)\, e^{-\alpha(r_L-r-1)}. \tag{7.18}$$

We can use Eqs. (7.15) and (7.3) to derive the values of $F_{mo} - rF'_{mo}$ for the exponential on-ramp and off-ramp. Performing the necessary algebra, we obtain

$$F_{mo}(r, r_L) - rF'_{mo}(r, r_L) = \begin{cases} 1 - h(\alpha)\left[\frac{1+(\alpha r)^3}{1+\alpha r}\right] e^{-\alpha(r_L - r - 1)}, & \text{exp-on} \\ h(\alpha)\left[\frac{1+(\alpha r)^3}{1+\alpha r}\right] e^{-\alpha(r_L - r - 1)}, & \text{exp-off} \end{cases} \tag{7.19}$$

where $F'_{mo}(r, r_L) = \partial F_{mo}(r, r_L)/\partial r$ as usual. These expressions are similar to the ones that we obtained for the LQ current profiles in Eq. (7.7). In particular, $F_{mo} - rF'_{mo}$ is a positive quantity for LQ-off and exp-off, and it is 1 minus that positive quantity for LQ-on and exp-on. Consequently, $F_{mo} - rF'_{mo}$ cannot vanish for the exponential off-ramp, and the neutral points must be confined to the leading sheath. But $F_{mo} - rF'_{mo}$ can vanish for the on-ramp, which means that one neutral point will move into this region after its birth at the $1o/mo$ boundary.

Figure 7.7 shows the space-time maps for the exponential on-ramp (left) and off-ramp (right). As we supposed, these maps are similar to the space-time maps for the LQ current profiles in the lower panels of Fig. 7.2. For each on-ramp, the neutral points are born at $r = r_L - 2$. The sinks of flux drift an almost imperceptibly small distance into the $1o$ sheath, toward a location close to the trailing edge of the sheath. The sources move gradually across the middle region. For the off-ramp, the neutral points form inside the $1o$ sheath and then move in opposite directions toward the front and back of the sheath.

For the on-ramp, the birth of the neutral points is easily obtained by setting $F_{mo} - rF'_{mo} = 0$ with $r = r_L - 2$. The result is a quadratic equation whose solution is

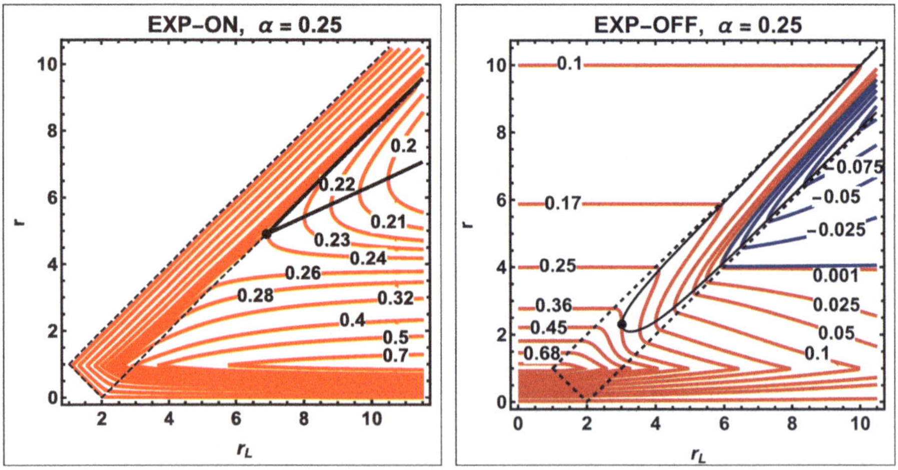

Fig. 7.7 Contours of constant flux for the exponential on-ramp (left) and off-ramp (right) with a decay constant $\alpha = 0.25$. Dotted boxes indicate the boundaries of the leading sheath. The black curves show the *loci* of neutral points where the curves have vertical slopes, and the black dot separates these curves into a lower segment where flux emerges and an upper segment where flux is annihilated. Red and blue colors indicate field lines whose circulation is clockwise and counterclockwise, respectively (when $y > 0$)

$\alpha r = (1 + \sqrt{1 - 4[1 - e^{\alpha}/h(\alpha)]})/2$. For $\alpha \ll 1$, we can expand this expression and the corresponding flux fraction, $F_{mo}(r, r_L)/r$ in powers of α to obtain

$$r = \frac{1}{\alpha} + 1 - \frac{3}{5}\alpha + \frac{19}{15}\alpha^2, \tag{7.20a}$$

$$r_L = \frac{1}{\alpha} + 3 - \frac{3}{5}\alpha + \frac{19}{15}\alpha^2, \tag{7.20b}$$

$$\Phi/\Phi_0 = \alpha - \alpha^3. \tag{7.20c}$$

A comparison with Eqs. (7.9)–(7.11) indicates that this exponential on-ramp falls in the category of ramps like LQ-on and QL-off, whose neutral-point coordinates vary inversely with the slope of the ramp, rather than inversely as the square root of the slope that we found for the QL on-ramp and LQ off-ramp. Likewise, the "falling distance", $r - (\Phi_0/\Phi) \approx 1 - (8/15)(3\alpha)$ for the EXP on-ramp. This supports our expectations that this exponential on-ramp would give fields similar to those of the LQ on-ramp, and suggests that when we perform a similar calculation for the exponential off-ramp, we will obtain coordinates that vary as the square root of the slope.

After its birth at r_L, the neutral point will move across the middle region (mo) on a path that can be derived by solving $F_{mo} - rF'_{mo} = 0$ without an auxiliary condition. For large r, we obtain $r_L \approx r + 1 + (2/\alpha)\, ln(\alpha r)$, which describes the gradual spreading of the source region in the left panel of Fig. 7.7. (Also, this relation shows that the lower branch of the neutral-point locus is not always linear as it appears in this figure, but develops some logarithmic curvature when r_L lies outside the field of view.)

For the off-ramp, the birth of the neutral points can be found by solving $F_{1o} - rF'_{1o} = 0$ and $F''_{1o} = 0$ simultaneously with F_{1o} given by Eq. (7.15). For a given value of α, these equations can be solved numerically for r and r_L. Algebraically, we obtain a complicated pair of transcendental equations in the variables $x = r_L - r - 1$ (which we suppose is small) and r. However, an approximate solution can be obtained by expressing each of these variables as power series in α and solving for the coefficients that make the resulting series expansions for F''_{1o} and $F_{1o} - rF'_{1o}$ begin at arbitrarily high powers of α. As we expected, the solution involves half-integral powers of the expansion variable, which in this case is 3α, analogous to $3/\Delta r$ for the linear ramps. Retaining terms of order $\alpha^{3/2}$ for x and α^1 for r, we obtain

$$x = \frac{1}{4}(3\alpha)^{1/2} + \frac{5}{24}(3\alpha) - \frac{1}{9}(3\alpha)^{3/2} \tag{7.21a}$$

$$r = 2(3\alpha)^{-1/2} - \frac{1}{3} + \frac{3}{8}(3\alpha)^{1/2} + \frac{5}{54}(3\alpha), \tag{7.21b}$$

so that the coordinates of the neutral points and the corresponding flux fraction are

$$r = 2(3\alpha)^{-1/2} - \frac{1}{3} + \frac{3}{8}(3\alpha)^{1/2} + \frac{5}{54}(3\alpha) \tag{7.22a}$$

$$r_L = 1 + 2(3\alpha)^{-1/2} - \frac{1}{3} + \frac{1}{8}(3\alpha)^{1/2} - \frac{25}{216}(3\alpha), \tag{7.22b}$$

$$\Phi/\Phi_0 = \frac{1}{2}(3\alpha)^{1/2} - \frac{1}{12}(3\alpha) - \frac{7}{144}(3\alpha)^{3/2}. \tag{7.22c}$$

These expressions are similar to those obtained for the LQ off-ramp, as well as the QL on-ramp and the linear on- and off-ramps, and the "falling distance", $\Phi_0/\Phi - r \approx (2/3) - (1/8)(3\alpha)^{1/2}$, is essentially the same.

Thus, for small α (and a correspondingly large decay time, $1/\alpha$), the neutral point forms close to the middle of the $1o$ sheath at a distance of approximately $2(3\alpha)^{-1/2}$ from the center of the sphere. For the value of $\alpha = 0.25$, that was used in Fig. 7.7, Eqs. (7.22b) and (7.22a) give $r_L = 3.00$ and $r = 2.37$, respectively, which are comparable to the values $r_L = 3.02$ and $r = 2.30$, obtained from the more accurate numerical solution, and plotted in the right panel of Fig. 7.7. Likewise, Eq. (7.22c) gives $\Phi/\Phi_0 = 0.34$, which is comparable to the plotted value of $\Phi/\Phi_0 = 0.36$.

The exponential maps have a property not found in the other space-time maps. Outside the sphere, there is a special location, $r_e = 1/\alpha$, at which the field lines neither rise nor fall. In Fig. 7.7, this location occurs at $r_e = 4$. For the on-ramp, field lines move toward this position from above and below, whereas for the off-ramp, the field lines move away from this location, some moving inward to annihilate their counterparts inside the sphere, and others rising upward as part of the outgoing bubble.

Referring to the right panel of Fig. 7.7, we can see that the collapsing loops have relatively low speeds, at least compared to the outgoing waves of speed c. We can determine the ingoing speeds from the slopes of the red contours, which are obtained by setting $d\Phi/\Phi_0 = d(F_{mo}/r) = 0$. The result is

$$\frac{dr}{dr_L} = \left\{\frac{\partial(F_{mo}/r)}{\partial r_L}\right\}\left\{\frac{\partial(F_{mo}/r)}{\partial r}\right\}^{-1} = \frac{r\,(\partial F_{mo}/\partial r_L)}{F_{mo} - rF'_{mo}}, \tag{7.23}$$

where the prime refers to differentiation with respect to r. Substituting F_{mo} from Eq. (7.3) and $F_{mo} - rF'_{mo}$ from Eq. (7.19), we find

$$\frac{dr}{dr_L} = -(\alpha r)\left\{\frac{1-(\alpha r)^2}{1+(\alpha r)^3}\right\}. \tag{7.24}$$

For $\alpha r \gg 1$, the speed, dr/dr_L approaches 1 (equivalent to $dr/dt = c$) as one would expect for the outgoing wave. (Of course, these results refer to the mo region (where $r < r_L - 2$), so when we allow r to become large, we must also allow r_L to become large to stay in this region.) For $\alpha r \ll 1$, $dr/dr_L \approx -\alpha r$ (equivalent to $dr/dt \approx -\alpha(r/R)c$. In particular, for $r = R$, the speed is approximately $dr/dt = -\alpha c$, which is $-0.25c$ for the red contours in the right panel of Fig. 7.7.

The special surface at $r = r_e$ is a boundary separating the decaying dipole field from the outgoing bubble. Referring to Eq. (7.3), we see that F_{mo} changes sign at $r = r_e$. For $r \ll r_e$, $F_{mo} \approx e^{-\alpha(r_L-1)}$, corresponding to a decaying dipole field of unit strength. On the other hand, for $r \gg r_e$, $F_{mo} \approx -\alpha r\ e^{-\alpha(r_L-r-1)}$. Therefore, $F_{mo}/r \approx -\alpha\ e^{-\alpha(r_L-r-1)}$, which is the amount of flux between the radius, r, in the middle region and the leading edge of the disturbance at $r = r_L$. Setting $r = r_L - 2$, we obtain $F_{mo}/r = -\alpha\ e^{-\alpha} \approx -\alpha$, when $\alpha \ll 1$. In this case, α is the amount of counterclockwise directed meridional flux in the leading sheath. Also, because $F_{mo}(r_e, r_L) = 0$, there is no net flux between $r = r_e$ and $r = r_L$. Thus, the amount of flux in the leading sheath returns across the progressively widening region between $r = r_e$ and $r = r_L - 2$.

Finally, referring to F_{mi} in Eqs. (7.14) and (7.17), we find a uniform field inside the sphere, growing to unit strength (for the on-ramp) and decaying to zero from unit strength (for the off-ramp).

Figure 7.8 shows evolving field lines corresponding to the space-time map for the exponential off-map with $\alpha = 0.25$ in Fig. 7.7. The left panels show the Pinchoff of the last few red field lines of clockwise meridional circulation. The right panels show the subsequent collapse of the inward components of each Pinchoff as they move toward the sphere and eventually annihilate their counterparts from within the sphere. The result is a weaker uniform field inside the sphere and a correspondingly weaker extension of this field outside the sphere, as indicated in the last panel at $r_L = 8.5$.

By $r_L = 6.1$, the original dipole field lines have become detached from the sphere, and the emergence of the blue bubble of oppositely directed field is visible in the lower left panel. Its trailing edge remains at $r = r_e = 4$, but its leading edge moves outward toward the front of the sheath where it will eventually replace the red field lines being annihilated there. At the back of the sheath, the O-type neutral point is an ongoing source of new contours of blue flux. The result is a collapsing dipole of clockwise directed (red) flux inside $r = r_e = 4$ and an outgoing bubble of counterclockwise directed (blue) meridional flux, concentrated in the sheath, but spread over the gradually increasing distance between $r = 4$ and the back edge of the outgoing sheath. The asymmetry produced by the LQ off-ramp is similar to the asymmetry of this exponential bubble, except that the returning flux for the LQ off-ramp is distributed over the finite distance, Δr, between $r = r_L - 2$ and $r = r_L - \Delta r - 2$.

Figure 7.9 provides an overview of the field evolution for a full range of time that shows the birth of the neutral points as well as the appearance and development of the bubble of blue flux. However, in this case, we use $\alpha = 1/3$ and show the entire field of view, scaled to the leading edge of the disturbance. This overview shows multiple phases of the evolution: At $r_L = 1.66$, the first wave (dashed circle) has left the sphere, but the second wave (also dashed) is still moving inward toward the center. By $r_L = 2.65$ the second wave has reached the center, and is heading outward again, soon to join the first wave to define the sheath. At this time, the field is stretched out and the neutral points are about to form.

By $r_L = 4.63$, red field lines are disappearing at the O-type neutral point at the front of the sheath and emerging at the X-type neutral point at rear of the sheath.

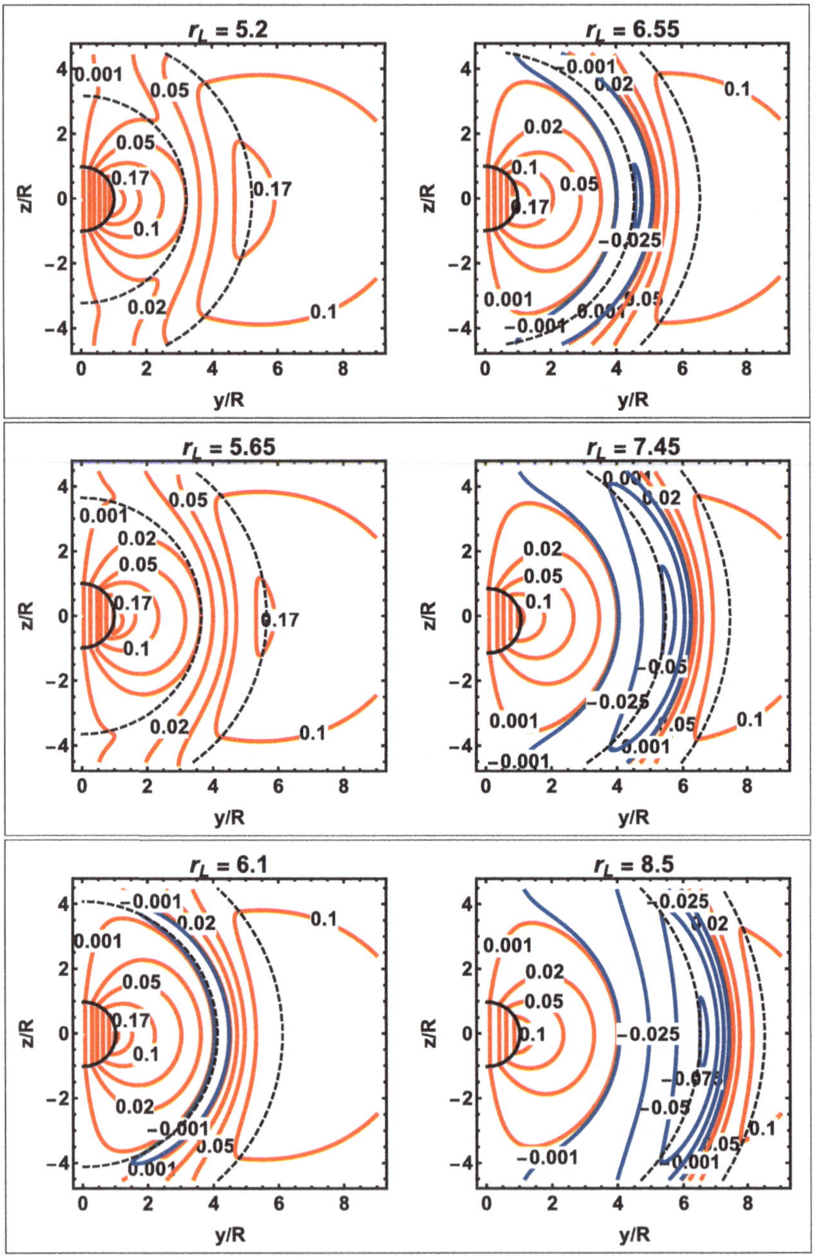

Fig. 7.8 Field lines for an exponential off-ramp of decay rate $\alpha = 0.25$. (Left) The end of the separation phase of red field lines, and the emergence of blue, reversed-polarity field. (Right) The continued emergence of blue flux near the back of the sheath (dashed lines) and the disappearance of red flux toward the front of the sheath. The bubble's trailing edge is fixed at $r = 1/\alpha = 4$, the top of the decaying dipole field

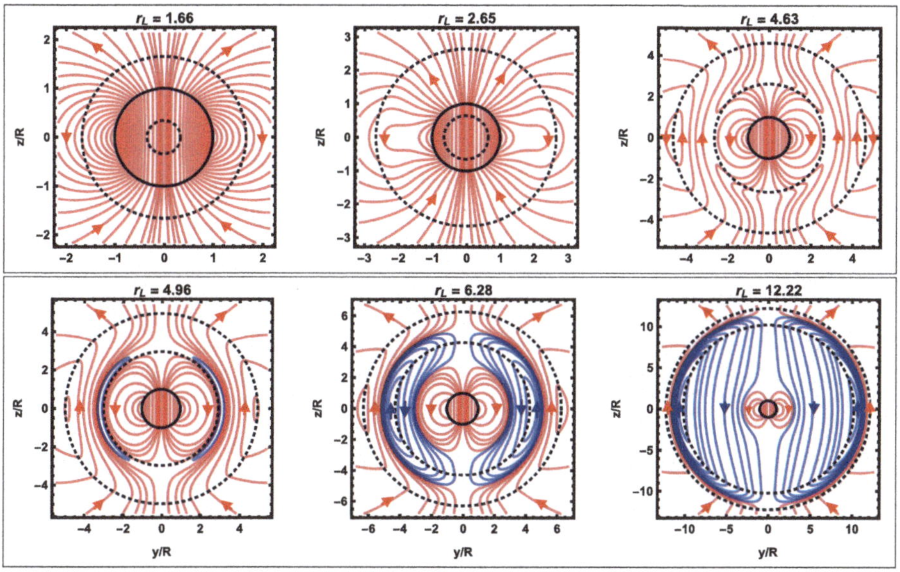

Fig. 7.9 Field lines produced by turning an initial current off gradually with the profile $G(\xi) = e^{-\alpha\xi}$, where the decay time, $1/\alpha$, is 3 units of R/c and ξ is the elapsed time since the start of the off-ramp. The outgoing wave fronts are indicated by dotted circles, and the spatial scale is varied to follow the first of these wave fronts at the leading edge of the outgoing disturbance

By $r_L = 4.96$, the Pinchoff is complete and a slice of blue field has emerged. By $r_L = 6.28$, the front of the bubble has expanded into the sheath, but its back end remains fixed at $r = r_e = 3$. New flux continues to emerge at the O-type neutral point, located at the trailing edge of the sheath. Finally, at $r_L = 12.22$, the field has evolved into a decaying dipole inside $r = r_e$ and an increasingly asymmetric bubble of outgoing flux.

Figure 7.10 is a similar overview for the corresponding exponential on-ramp again with $\alpha = 1/3$, shown here for completeness. The initially linked field lines gradually pinch off to separate the outgoing bubble from the growing dipole field. If we set $F_{mo} - rF'_{mo} = 0$ in Eq. (7.19) and assume that $\alpha r \gg 1$, we obtain $r_L - 1 - r \approx (2/\alpha)ln(\alpha r)$ for the relation between the elapsed time, r_L, and the radial position, r, of the neutral point. This means that the return path of the bubble's flux gets increasingly wider with time, while the neutral point moves continuously outward in front of the growing loops of the dipole field.

7.4 Symmetric Cubic Profiles

If the current ramp were curved at both ends as shown for the symmetric cubic profiles in Fig. 7.11, then we would not expect to find sharp boundaries at either end of the field line wave train or its space-time display. We can verify this expectation

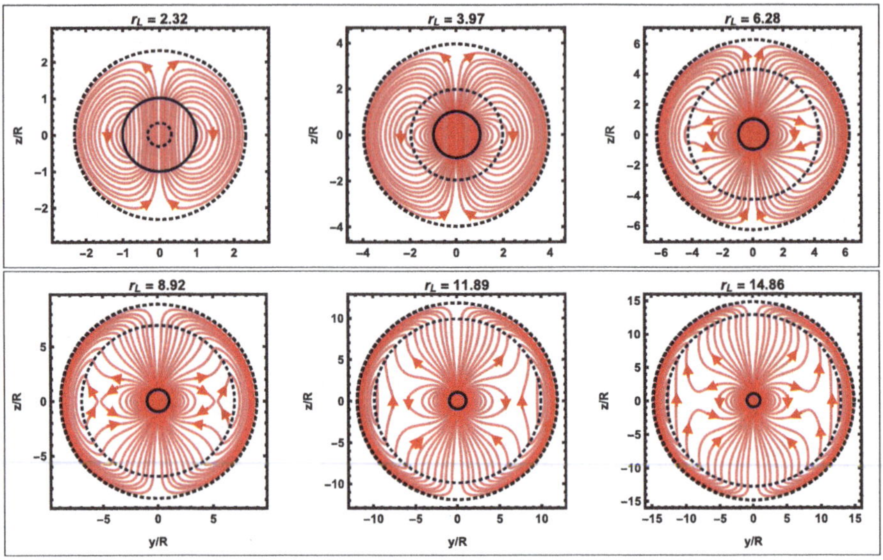

Fig. 7.10 Field lines produced by turning the current on gradually with the exponential profile $G(\xi) = 1 - e^{-\alpha\xi}$, where the rise time, $1/\alpha$, is 3 units of R/c and ξ is the elapsed time since the start of the current ramp. Ingoing and outgoing wave fronts are indicated by dotted circles, and the spatial scale is varied to follow the leading edge of the outgoing disturbance

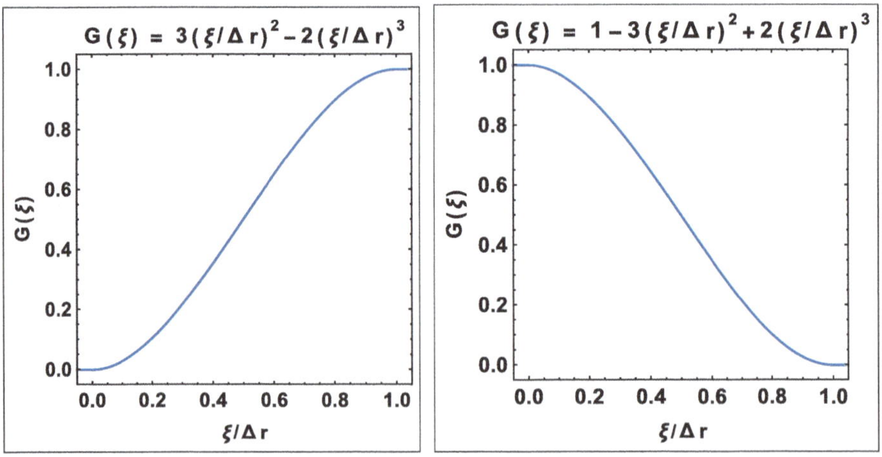

Fig. 7.11 Symmetrically shaped cubic profiles with the on-ramp on the left and off-ramp on the right

by comparing the "cubic" space-time maps in Fig. 7.12 with the "linear" ones in Fig. 6.4. Going from the left panel of Fig. 6.4 to the left panel of Fig. 7.12, we can see that the upper bends of the switchbacks lie at the front of the trailing sheath for the linear on-ramp, but the bends shift into the middle region (mo) for the cubic

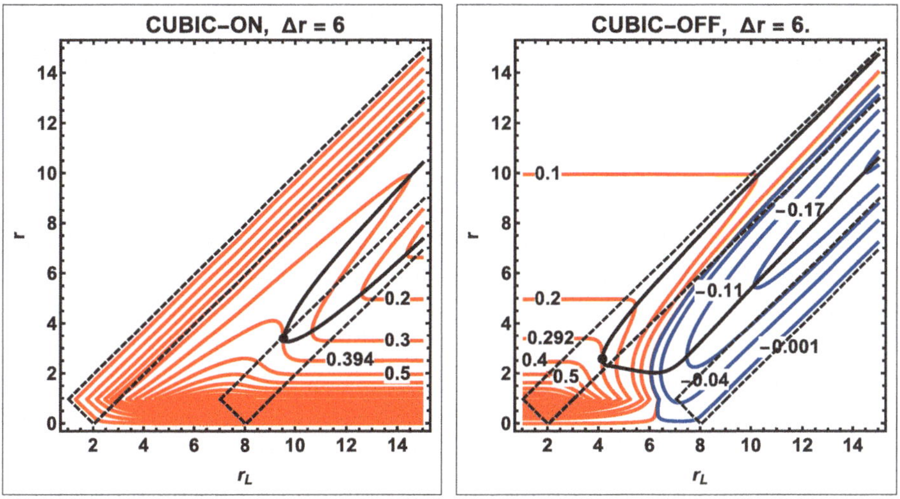

Fig. 7.12 Contours of constant flux for the symmetric cubic on-ramp (left) and off-ramp (right) with a width $\Delta r = 6$. Dotted boxes indicate the boundaries of the leading and trailing sheaths. The black curves show the *loci* of neutral points where the curves have vertical slopes. The black dot separates these curves into a lower segment where flux emerges and an upper segment where flux is annihilated. Red and blue colors indicate field lines whose circulation is clockwise and counterclockwise, respectively (when $y > 0$)

on-ramp. Likewise, for the off-ramps (right panels), the sources of the blue bubbles shift from the front of the trailing sheath to the center of the middle region. Thus, by changing from a linear profile to a cubic profile, we cause the bubbles to spread into the middle region and their O-type neutral points to become centered there.

It is instructive to show that the O-type neutral points asymptotically approach the midpoint of the middle region (mo) for the cubic on-ramp and off-ramp. For this purpose, we return to Eq. (5.7) and replace the quantity $dr'_L/\Delta r$ by $G'(r_L - r'_L)dr'_L$, with G' obtained from the left panel of Fig. 7.11 for the cubic on-ramp. After some algebra, we obtain

$$F_{mo} \; = \; -\frac{2}{(\Delta r)^3}\left(2r^3 - 3r_m r^2 + r_m^3\right) \; + \; A, \tag{7.25}$$

where r_m is the midpoint of the middle (mo) region, defined by $r_m = r_L - \Delta r/2 - 1$, and A represents the remaining lower-order terms given by $A = \frac{1}{2} + \frac{3}{2}\frac{r_m}{\Delta r}(1 - \frac{4}{5}\frac{1}{(\Delta r)^2}) \approx \frac{1}{2} + \frac{3}{2}\frac{r_m}{\Delta r}$ for $\Delta r > 2$.

From Eq. (7.25), we can calculate the value of $F_{mo} - rF'_{mo}$:

$$F_{mo} - rF'_{mo} \; = \; \frac{2}{(\Delta r)^3}(r - r_m)\left(4r^2 + r_m r + r_m^2\right) \; + \; A. \tag{7.26}$$

Setting $F_{mo} - rF'_{mo} = 0$, dividing the resulting equation by the positive, second-order quantity, $4r^2 + r_m r + r_m^2$, and expressing all lengths in units of Δr, we obtain

$$2(r - r_m) + \frac{1/2 + 3r_m/2}{4r^2 + r_m r + r_m^2} = 0. \tag{7.27}$$

If we let r and r_m become large while keeping $r - r_m$ finite, then $4r^2 + r_m r + r_m^2 \approx 6r_m^2$ and $r \approx r_m - 1/8r_m$ for large r_m (i.e., $r_m \gg \Delta r$ in unnormalized units). In this case, $r \to r_m$ and the O-type neutral point becomes asymptotically centered in the mo region.

For the off-ramp, the neutral point is obtained by setting the on-ramp value of $F_{mo} - rF'_{mo}$ equal to 1. In this case, we obtain

$$\frac{2}{(\Delta r)^3}(r - r_m)\left(4r^2 + r_m r + r_m^2\right) + A = 1. \tag{7.28}$$

It follows that

$$2(r - r_m) + \frac{-1/2 + 3r_m/2}{4r^2 + r_m r + r_m^2} = 0, \tag{7.29}$$

where all distances are expressed in units of Δr. For large r and r_m, this again leads to the result that $r \approx r_m - 1/8r_m$ and that $r \to r_m$ as r_m increases beyond Δr. In fact, the replacement of $+1/2$ in Eq. (7.27) by $-1/2$ in Eq. (7.29) causes the neutral point to be slightly closer to r_m for the cubic off-ramp than for the cubic on-ramp with the same value of r_L.

Figure 7.13 shows the field lines at the time $r_L = 21$ for the cubic on-ramp (left panel) and off-ramp (right panel). By this time, the outgoing bubbles have formed and the neutral points are very close to their asymptotic values. For the on-ramp in the left panel, the O-type neutral point is nicely centered between the two sheaths where clockwise-directed (red) flux is submerging. The X-type neutral point is located at the back of the trailing sheath where meridional flux is accumulating as the dipole field lines are being pinched off.

For the off-ramp in the right panel, the O-type neutral point is again centered between the two sheaths where counterclockwise directed (blue) flux is emerging. We can infer the presence of the other neutral point at the front of the leading edge by the surrounding red lines where clockwise-directed (red) flux is being removed to gradually equalize the amount of flux in the leading and trailing sheaths. And, of course, for the off-ramp, no field is left behind.

Next, we list the neutral-point coordinates and associated flux fractions for the cubic on- and off-ramps. Beginning with the on-ramp, we have

$$r = \frac{1}{6}\left(3 + \sqrt{6}\right) + \frac{1}{6}\sqrt{6}\,(1/\Delta r)^{-1} + 0.8696\,(1/\Delta r) - 4.4617\,(1/\Delta r)^2, \tag{7.30a}$$

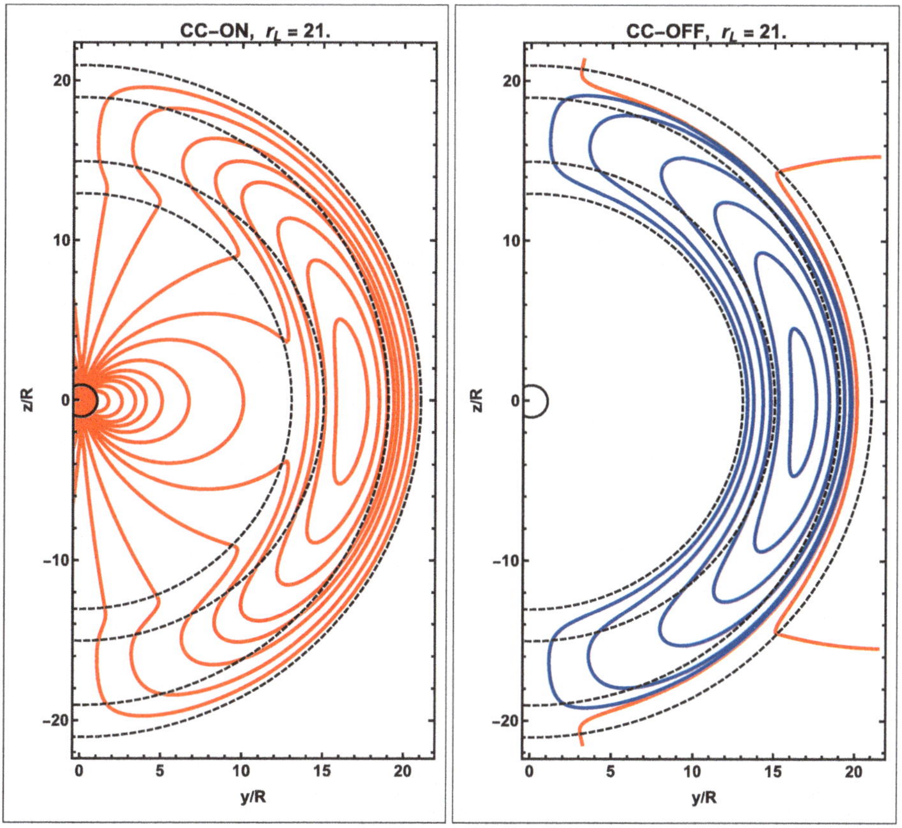

Fig. 7.13 Magnetic field lines for the symmetric cubic on-ramp (left) and off-ramp (right) of width $\Delta r = 6$, showing the asymptotic locations of the O-type neutral points in the centers of the outgoing bubbles. Dashed circles indicate the boundaries of the leading and trailing sheaths, and the spherical current source is indicated by the small circle of radius $r = 1$ centered at (0,0)

$$r_L = \frac{1}{6}\left(3+\sqrt{6}\right) + \frac{1}{6}\left(6+\sqrt{6}\right)(1/\Delta r)^{-1} + 0.5700\,(1/\Delta r)$$
$$+\,2.4821\,(1/\Delta r)^2, \tag{7.30b}$$

$$\Phi/\Phi_0 = \sqrt{6}\,(1/\Delta r) + \left(3-\sqrt{6}\right)(1/\Delta r)^2 - 7.3142\,(1/\Delta r)^3. \tag{7.30c}$$

For simplicity, we have replaced complicated factors of the form $(1/c)(a + b\sqrt{6})$ (where a, b, and c are large integers) by their numerical equivalents. A similar treatment for the cubic off-ramp gives

$$r = \frac{1}{6}\left(-3+\sqrt{6}\right) + \frac{1}{6}\sqrt{6}\,(1/\Delta r)^{-1} + 1.4030\,(1/\Delta r)$$

$$+\ 2.3827\,(1/\Delta r)^2, \tag{7.31a}$$

$$r_L = \frac{1}{6}\left(9+\sqrt{6}\right) + \frac{1}{6}\left(6+\sqrt{6}\right)(1/\Delta r)^{-1} - 1.5634\,(1/\Delta r)$$
$$+\ 0.8821\,(1/\Delta r)^2, \tag{7.31b}$$

$$\Phi/\Phi_0 = \sqrt{6}\,(1/\Delta r) - (3+\sqrt{6})\,(1/\Delta r)^2 + 7.8858\,(1/\Delta r)^3. \tag{7.31c}$$

For $\Delta r \gg 1$, the leading non-constant term of the series for r varies linearly with Δr as $r \sim \Delta r/\sqrt{6}$, and the leading term of the series for the flux fraction has the reciprocal variation, $\Phi/\Phi_0 \sim \sqrt{6}/\Delta r$. Thus, the symmetric cubic profiles fall in the category of the QL off-ramps and the LQ on-ramps, whose neutral points form in the curved portion of the ramp. The "falling distances" are $1 - 0.370/\Delta r$ for the cubic on-ramp and $1 - 0.697/\Delta r$ for the cubic off-ramp.

Generalizing from all of our examples (linear ramps, quadratic ramps, exponential ramps, and the cubic ramps), the square-root dependence occurs only when the neutral points are born at the linear end of the current ramp; otherwise, the dependence on ramp width, Δr, or slope (in the case of the exponentials) is linear. Moreover, when the square root does occur, its argument is 3 times the slope of the linear segment of the ramp.

In Fig. 7.13, the symmetric, outgoing bubbles remind us of the bubbles produced by the sudden on-ramp and off-ramp, shown previously in Figs. 3.1 and 4.1. The difference is that the bubbles produced by the cubic ramps are bounded by leading and trailing sheaths, whereas the sudden current steps do not produce such sheaths. In the next subsection, we shall reexamine the fields of the sudden turn-on and turn-off more closely using space-time maps and field-line drawings.

7.5 Sudden Profiles

Figure 7.14 is the space-time map for the sudden turn-on whose field lines for $r_L =$ 1.3–4.0 were shown in Fig. 3.1 This enlarged map allows us to see the detailed behavior around the time that the neutral point was born. Its origin at $r = 1$ occurred when $r_L = 1 + 6^{1/3} = 2.81712$, as indicated by the black dot. The subsequent motion of this neutral point is indicated by the black curve, which intersects the red contours of constant flux where their slopes are infinite, beginning with the track at $\Phi/\Phi_0 = f(1, 1 + 6^{1|3}) = (3/2)(6^{1/3} - 1) \approx 1.22568$.

Like the neutral-point curves in previous space-time maps, this curve intersects the space-time tracks at points that the tracks come in from the left in a concave geometry where flux is being removed. However, unlike the previous neutral-point curves, this neutral-point curve does not have a lower branch where flux emerges. Instead, the flux emerges at the trailing edge of the disturbance (dashed curve) where the slopes of the tracks have corner-like discontinuities. The "pair production" that occurs here sends part of the flux outward to merge and disappear at the neutral-

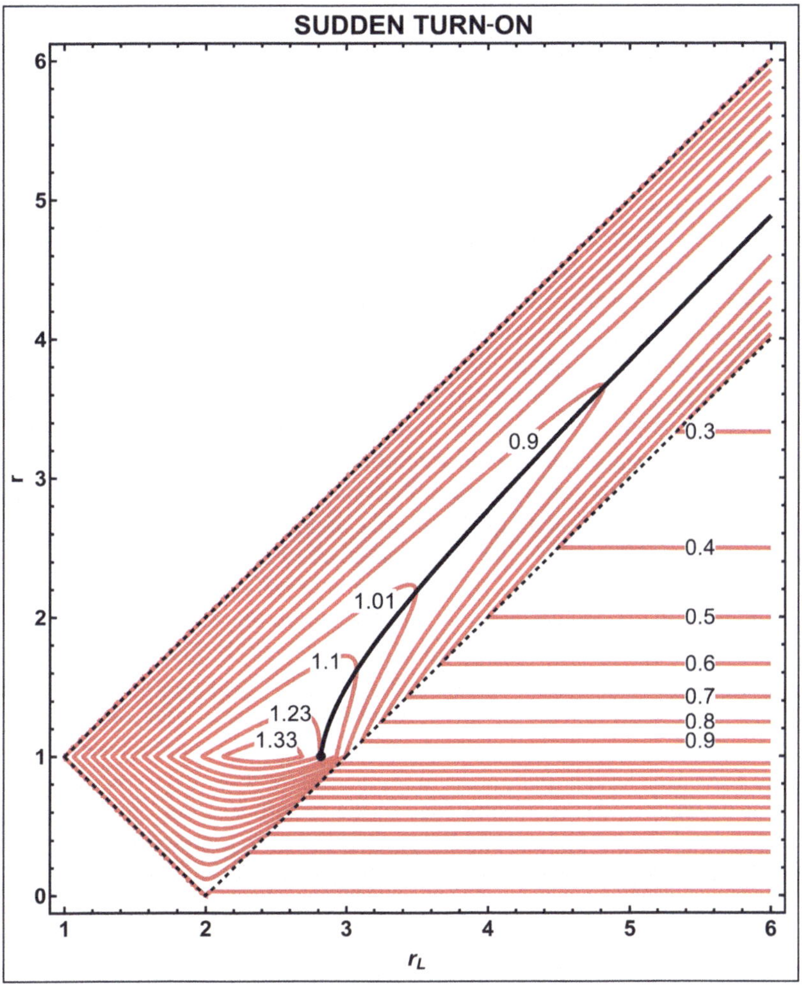

Fig. 7.14 Space-time tracks obtained from contours of constant flux for a sudden step up of current, showing the evolution of the equatorial field lines. The solid black curve is the *locus* of neutral points where the tracks have vertical slope, and the black dot indicates their birth place at the surface of the sphere. Dashed lines indicate the wave fronts that move inward and outward from the surface of the sphere. The neutral point eventually becomes centered in the outgoing bubble of positive flux

point curve; the other part is left in place as a static loop of the remaining dipole field.

Beginning at $r_L = 1$ many tracks diverge from $r = 1$. The earlier ones go to great heights and return their flux only via the exchange process described above. The later tracks do not reach such great heights and therefore fall back to the surface where they meet their companion from below. The field line marked 1.23 is the boundary between these two processes. Contours of greater flux, like the contour marked 1.33 in Fig. 7.14, return to the surface, whereas contours of less flux, like the one marked 0.9 in Fig. 7.14, return their flux via the exchange process.

Figure 7.15 shows the corresponding field lines. In the left panels, a field line marked 1.25 annihilates its counterpart at the surface during $r_L = 2.75$–2.8. In the right panels, a field line marked 1.15 slips across the surface to form the back end of a shrinking loop at the O-type neutral point. In both of these cases, extra flux with $\Phi/\Phi_0 > 1.0$ has been removed from the sphere prior to the arrival of the trailing edge of the disturbance (dashed line).

This flux removal process stops when the trailing edge of the disturbance crosses the sphere. As shown in Fig. 7.14, only the horizontal contours of the static dipole field remain after the passage of the trailing edge. Figure 7.16 shows the field lines after the wave has moved out from the sphere. In the left panels, the field line marked 0.9 pinches off at the location of the trailing edge. Its inward component just sits there as part of the static dipole field extending from its uniform counterpart inside the sphere. The outward component becomes part of the field that circulates around the O-type neutral point and gradually shrinks away, as seen in the right panels. By this time ($r_L = 3.325$–3.425), the field line marked 0.75 has pinched off, adding flux to the static dipole as well as the outgoing bubble.

In summary, the sudden turn-on of current causes the flux inside the sphere to be temporarily larger than its final value of $\Phi/\Phi_0 = 1$. This excess amount is approximately 1.41 and occurs when $r_L \approx 2.41$. Flux contours with Φ/Φ_0 in the range 1.22–1.41 are removed at the surface of the sphere where they meet and annihilate collapsing loops from outside the sphere. Flux contours with Φ/Φ_0 in the range 1.00–1.22 simply pop out of the sphere to form the back ends of shrinking loops. For times $r_L \geq 3$, the trailing edge of the disturbance lies outside the sphere where contours with Φ/Φ_0 less than 1.0 pinch together to produce loops of the static dipole field and the back ends of shrinking loops in the outgoing bubble.

Figure 7.17 shows the space-time map for the sudden turn-off whose field lines during the interval $r_L = 1$–4 were shown in Fig. 4.1. Again, we use an enlarged version of this figure to show the birth and early evolution of the neutral point inside the sphere. A close examination of Fig. 7.17 reveals that the first few tracks have positive slopes in the region between the initially ingoing and outgoing waves (dashed lines). This means that after the passage of the ingoing wave, the uniform field lines slip across the surface of the sphere and annihilate their counterparts in the external dipole field.

The neutral points are born at $r_L = 1.25$ and $r = 0.75$, indicated by the black dot on the field line marked 0.562. After this birth, the space-time tracks bend forward from the neutral-point curve, with the convex geometry characteristic of a source of emerging flux. So this neutral-point curve is like the lower branch of the U-shaped curves that we obtained for the current ramps of finite duration. Part of this emerging flux moves outward across the surface of the sphere to annihilate dipole field lines at the leading edge of the disturbance. The other part moves inward to remove the remaining uniform field lines inside the sphere.

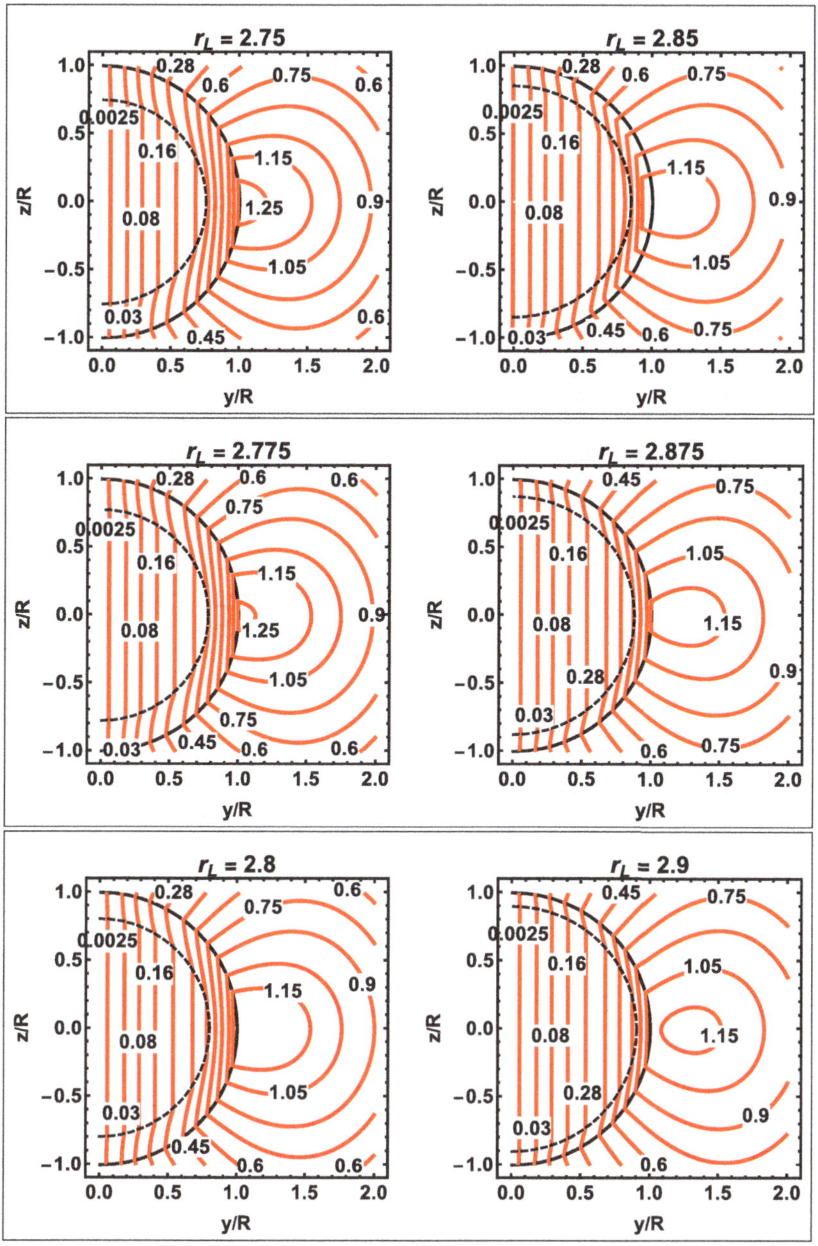

Fig. 7.15 Field lines for a sudden turn-on, showing two ways that strong fields with $\Phi/\Phi_0 > 1$ are shed from the sphere. (Left column) $r_L = 2.75$–2.8, showing the collapse and annihilation of the field line with $\Phi = 1.25$. (Right column) $r_L = 2.85$–2.9, showing the expulsion and shrinkage of the loop with $\Phi = 1.15$

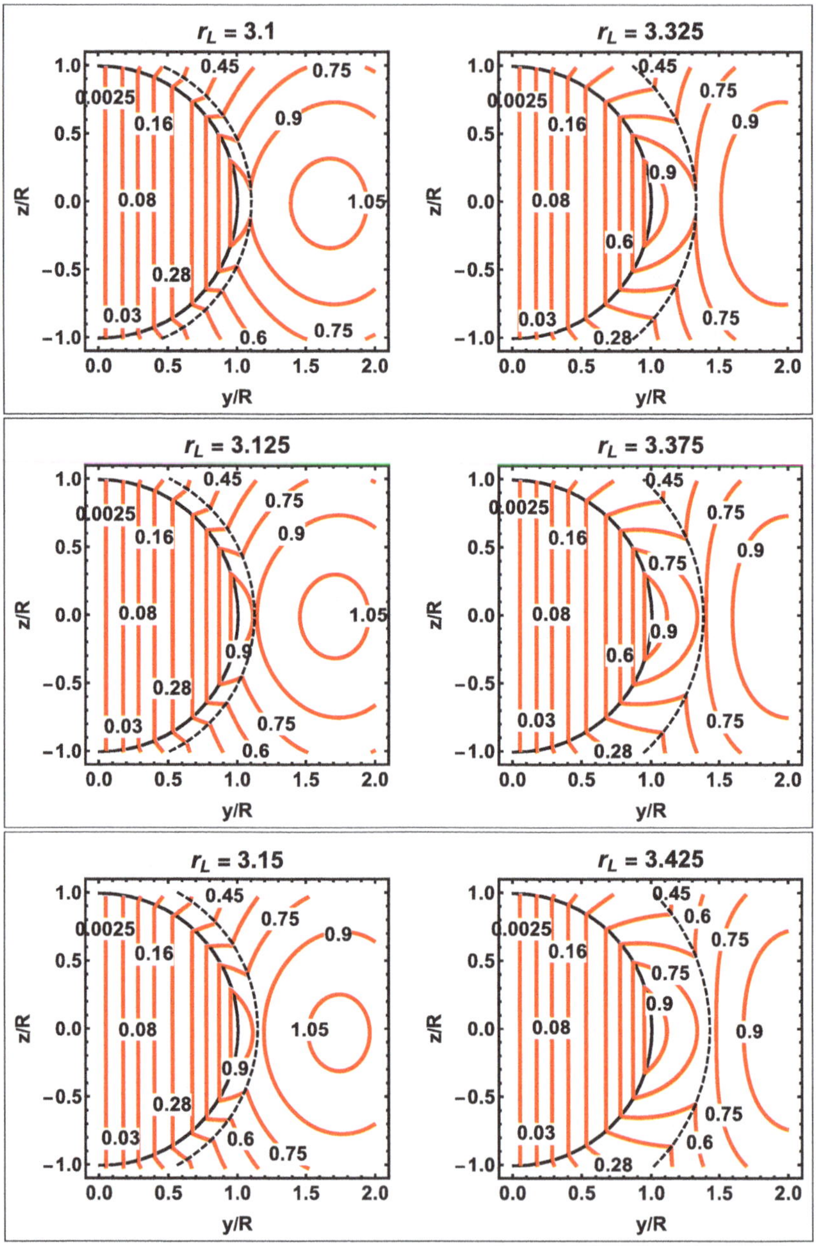

Fig. 7.16 A continuation of the previous figure, showing the detachment that occurs when the trailing wave (dotted circle) goes by. (Left column) $r_L = 3.1$–3.15, showing the detachment of the field line with $\Phi = 0.9$. (Right column) $r_L = 3.325$–3.425, showing the detachment of the field line with $\Phi = 0.75$ and the continued shrinkage of the detached loop with $\Phi = 0.9$ as it pulls ahead of the trailing wave

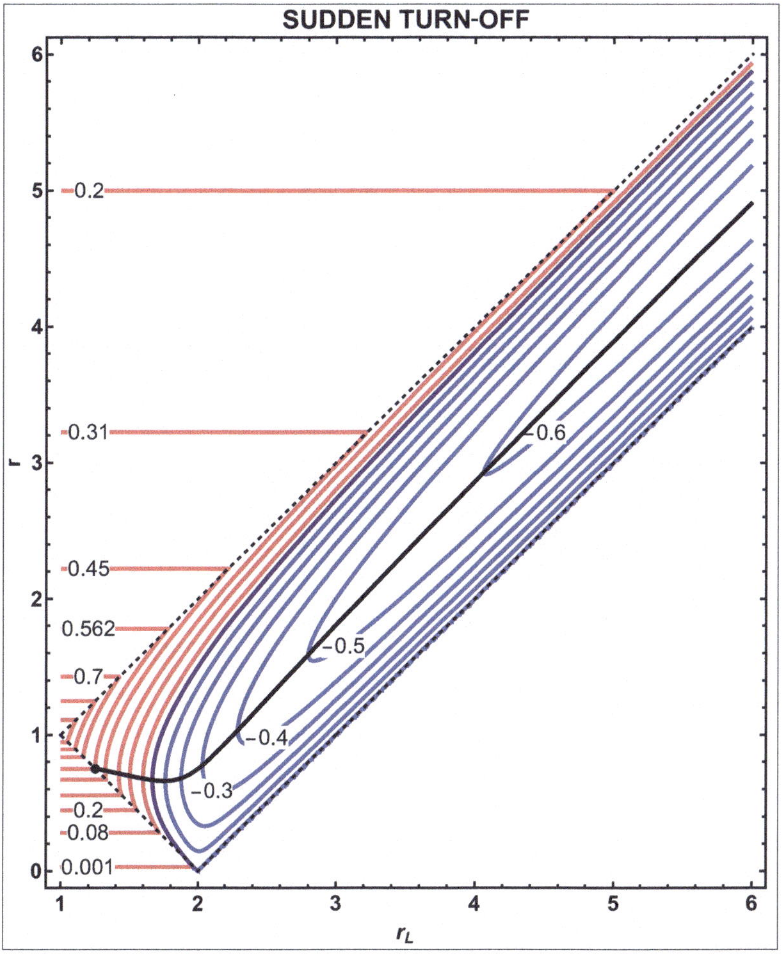

Fig. 7.17 Space-time tracks obtained from contours of constant flux for a sudden decrease of current, showing the evolution of the equatorial field lines. The solid black curve is the *locus* of neutral points where the tracks have vertical slope, and the black dot indicates their birth place inside the sphere. Dashed lines indicate the wave fronts that move inward and outward from the surface of the sphere. The neutral point eventually becomes centered in the outgoing bubble of negative flux

Further inspection of Fig. 7.17 shows that the pinchoff has been completed by the time that the neutral-point curve intersects the curve of zero flux (i.e., when $f = 0$). This occurs when $r_L = 5/3 \approx 1.67$ and $r = 2/3 \approx 0.67$, obtained by solving $f - rf' = 0$ and $f = 0$ with f given by $f = r^3 - f_i(r, r_L)$. After this time, reversed-polarity flux emerges at the neutral point.

As indicated by the black curve, this emergence begins inside the sphere and continues as this bubble of reversed-polarity flux moves outside and away from the sphere. The asymptotically centered location of this neutral point is similar to what

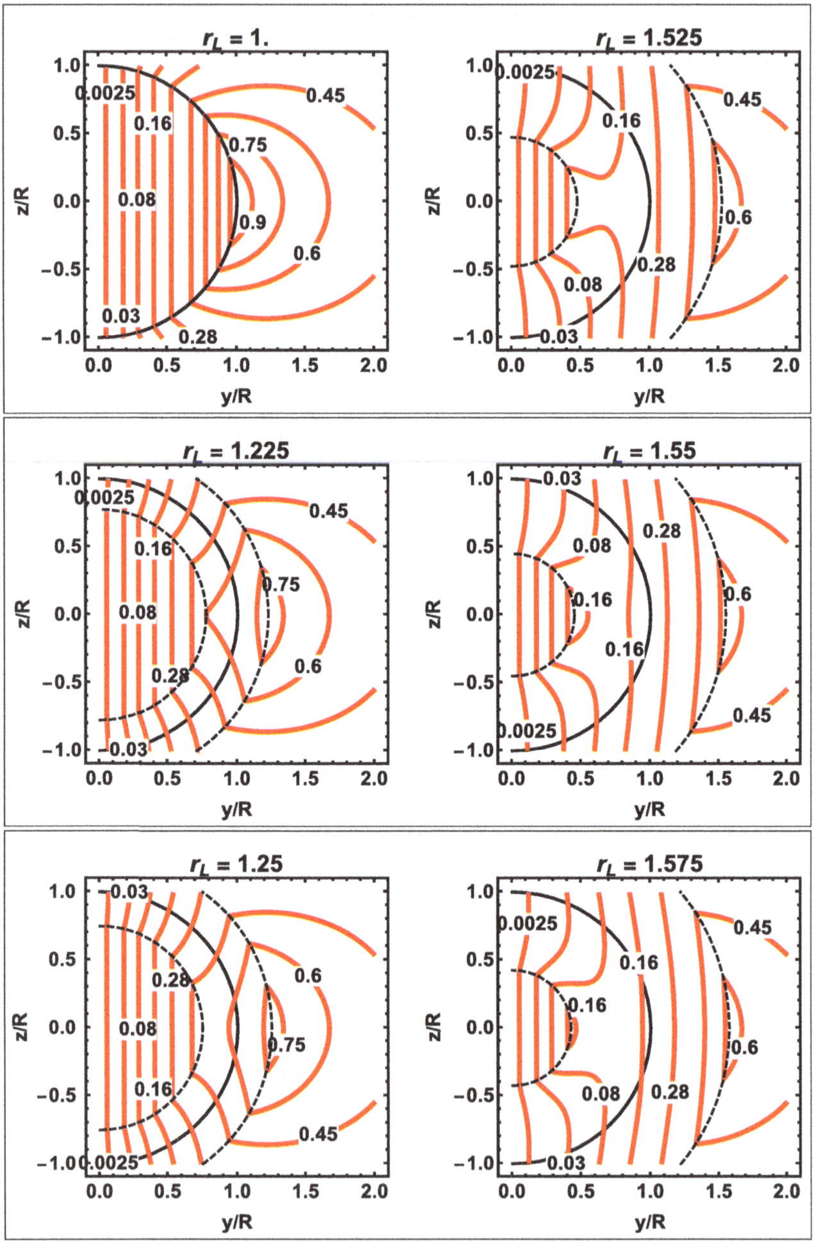

Fig. 7.18 Field lines for a sudden turn-off, showing the removal of uniform magnetic field lines from the sphere as the ingoing wave passes by. (Left column) $r_L = 1.0$–1.25, showing the original dipole field and the extraction of the uniform field line (labeled 0.6) inside the sphere. (Right column) $r_L = 1.525$–1.575, showing the Pinchoff of a field line (labeled 0.16) after the neutral point forms at $(r_L, r) = (1.25, 0.75)$

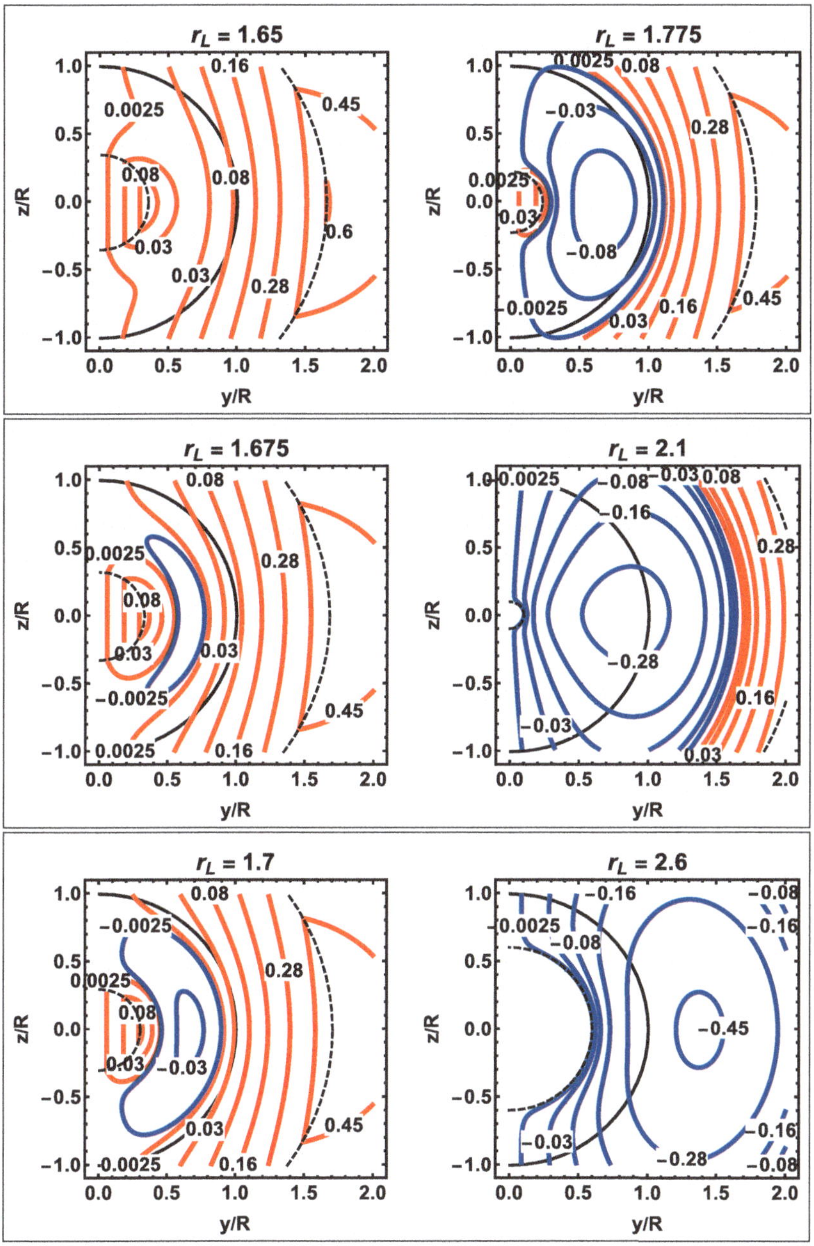

Fig. 7.19 A continuation of the previous figure, showing the final detachment of field lines inside the sphere from those outside, and the emergence of the reversed-polarity field. (Left column) $r_L = 1.65$–1.7, showing the detachment of the 0.0025 field line and the emergence of the reversed-polarity field lines. (Right column) $r_L = 1.775$–2.6, showing the shrinkage and annihilation of the original red field lines together with the growth and expansion of the blue bubble of flux as it approaches the center, reverses its direction, and heads outward away from the sphere

we obtained for the symmetric cubic profile. As the reader may recall, this similarity provided the motivation for making these space-time maps for the sudden turn-on and turn-off.

Figure 7.18 shows the field lines for this sudden turn-off during $r_L = 1$–1.575 prior to and just after the birth of the neutral point at $r_L = 1.25$. The upper left panel shows the initial dipole field outside the sphere and the uniform field inside the sphere prior to turning the current off at $r_L = 1$. During $r_L = 1.225$–1.25, the dashed wave fronts diverge inward and outward from the surface of the sphere. Field lines like the one marked 0.6 slip across the surface and collapse around the outgoing wave. However, as seen in the right panels, after $r_L = 1.25$, the field lines inside the sphere begin to pinch off, producing field lines that collapse around equatorial points at both the ingoing and outgoing waves.

Figure 7.19 shows the continuation of this process during $r_L = 1.65$–2.6, as the Pinchoff ends and the original dipole field becomes separated into two parts. The left panels show the emergence of the reversed-polarity (blue) field lines at the same place that the original (red) field lines were separated. As the ingoing wave continues toward the center of the sphere, the new bubble expands both inward and outward, and the central region of old red flux dies away. As seen in the right panels, the trailing edge of the bubble moves with the wave, turning around at the center, leaving no field behind it and following the first wave outward from the sphere.

Chapter 8
Other Examples of Interacting Fields

In the previous chapters, we have seen numerous examples in which an outgoing wave of one polarity moves through a field of opposite polarity and causes a collapse of field lines followed by the formation of a neutral point and the subsequent separation of the two fields. For the gradual turn-on, this process occurred in the trailing sheath, and marked the departure of the bubble of flux from the dipole field that was left behind. For the gradual turn-off, the action occurred in the leading sheath where the outgoing bubble of opposite polarity became visible. Intuitively, this seems like a general process that occurs whenever a field of one polarity passes through a field of opposite polarity and gradually exceeds it in strength. In this chapter, we will pursue this idea by considering other examples of interacting magnetic fields.

8.1 The Gradual Turndown

Although it is difficult to find a solution to Eqs. (2.12) and (2.13) that is any simpler than the current ramps that we have discussed in the previous chapters, a wave that differs by a multiplicative constant is also a valid solution. Consequently, we start with a dipole field of amplitude B_0 and gradually reduce the current by a factor of α over an interval of Δr. In this case, the f-value is given by

$$F(r, r_L) = \begin{cases} r^3 - \alpha F_{in}, & r \leq 1 \\ 1 - \alpha F_{out}, & r \geq 1, \end{cases} \tag{8.1}$$

Electronic Supplementary Material The online version of this chapter (https://doi.org/10.1007/978-3-030-40264-8_8) contains supplementary material, which is available to authorized users.

N. R. Sheeley, Jr., *Transient Magnetic Fields*,
https://doi.org/10.1007/978-3-030-40264-8_8

where F_{in} and F_{out} are the f-values inside and outside the s phere as given by Eqs. (5.8) and (5.9), respectively.

Figure 8.1 shows the corresponding space-time map when $\alpha = 0.3$ and $\Delta r = 6$. It resembles the space-time map obtained for the full turn-off shown in the right panel of Fig. 6.4. However, unlike the tracks in Fig. 6.4, which move sharply down through the middle region to meet their counterparts inside the sphere, most of the tracks in Fig. 8.1 are intercepted by the trailing sheath, which carries them upward again before depositing them at locations that are 30% lower than their starting positions. Thus, the track marked 0.25 starts at $r = 4$ on the left edge of the map and ends at $r = 2.8$ on the right edge. It has a normal life, moving slowly inward through the leading sheath, and then through the central region of the transient before being

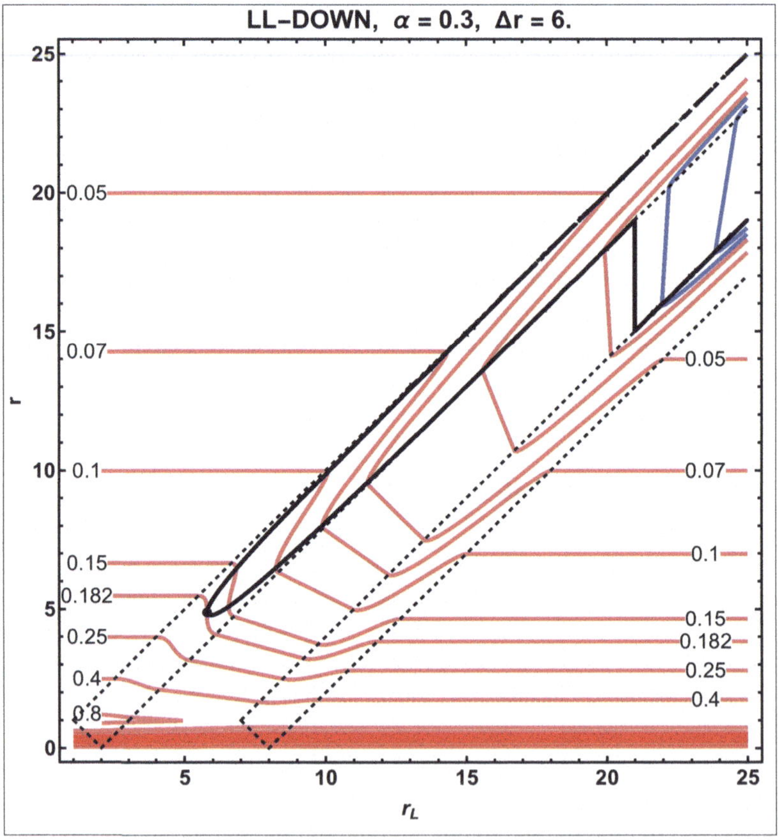

Fig. 8.1 Space-time tracks obtained when the current is linearly reduced by a factor of $\alpha = 0.3$ during a ramp time $\Delta r = 6$. The solid black curve is the *locus* of neutral points where the tracks have vertical slopes, and the black dot indicates their birth place. Dashed lines indicate the sheaths that are produced by the outgoing waves. Unlike the tracks for the full turn-off, which end as they meet their counterparts in the sphere (*cf.* LL-OFF in Fig. 6.4), these tracks do not end, but move to positions closer to the sphere

carried slightly upward as it passes through the trailing sheath. The track marked 0.182 is the last track to participate in this "normal" behavior. Tracks like the one marked 0.1 move backwards in time as they pass through the leading sheath.

The downward motion in the leading sheath is a consequence of the negative field of the turndown. Its negative field opposes the positive field of the dipole, producing a weaker net field whose field lines therefore separate and move inward. Eventually, the field acquires a neutral point where the legs of the distorted field lines pinch together and reconnect, producing a loop that moves inward to continue the "normal life" of the motion and the back end of a loop that rises rapidly upward to meet and annihilate its counterpart near the front of the leading sheath.

In the middle region, these contours of constant flux, Φ/Φ_0, have a speed, dr/dr_L, that can be derived from the flux relation

$$\Phi/\Phi_0 = F(r, r_L)/r = (1/r)\left[1 - \alpha\left(\frac{r_L - 1}{\Delta r}\right)\right]. \tag{8.2}$$

Taking the derivative while keeping Φ/Φ_0 constant, we obtain

$$dr/dr_L = -(1/\Phi)(\alpha/\Delta r), \tag{8.3}$$

with the flux expressed in units of Φ_0. From this relation, we can see that for contours of positive flux, the speed is inward toward the sphere and its magnitude is larger for contours with smaller fluxes. When the flux reaches 0, the speed becomes infinite and the black neutral-point curve drops vertically from the back of the leading sheath to the front of the trailing sheath. This happens when $r_L = 1 + \Delta r/\alpha = 21$, as one can see by setting the flux in Eq. (8.2) equal to zero. Beyond this point, the contours in this middle region are negative (blue) and the speeds are directed outward in the direction of the outgoing bubble.

Finally, as the positive meridional field in the trailing sheath arrives, the net field is strengthened and the field lines move outward again. At the back of the trailing sheath, the field lines are left as static loops 30% closer to the origin than they started when $r_L = 1$.

Figure 8.2 illustrates the evolution of the field lines whose space-time tracks along the "equator" were shown in the previous figure for the current turndown with $\alpha = 0.3$ and $\Delta r = 6$. These snapshots were obtained at times running top to bottom first in the left column and then in the right column. At $r_L = 5.62$, the last collapsing field line marked 0.182 shows its typical pinched shape within the leading sheath. By $r_L = 8.59$ that field line has been carried inward in the central region, and the line marked 0.1 has separated into two components—one that is beginning to move inward through the middle region and the other which is shrinking around the front of the leading sheath.

By $r_L = 10.57$, the 0.1 contour has nearly finished its passage through the middle region, and the 0.075 contour has divided and is starting its descent. By $r_L = 14.53$, the 0.1 contour has been carried back outward by the trailing sheath and the loop is about to be left in its final position. Meanwhile, the surviving 0.075 contour is being

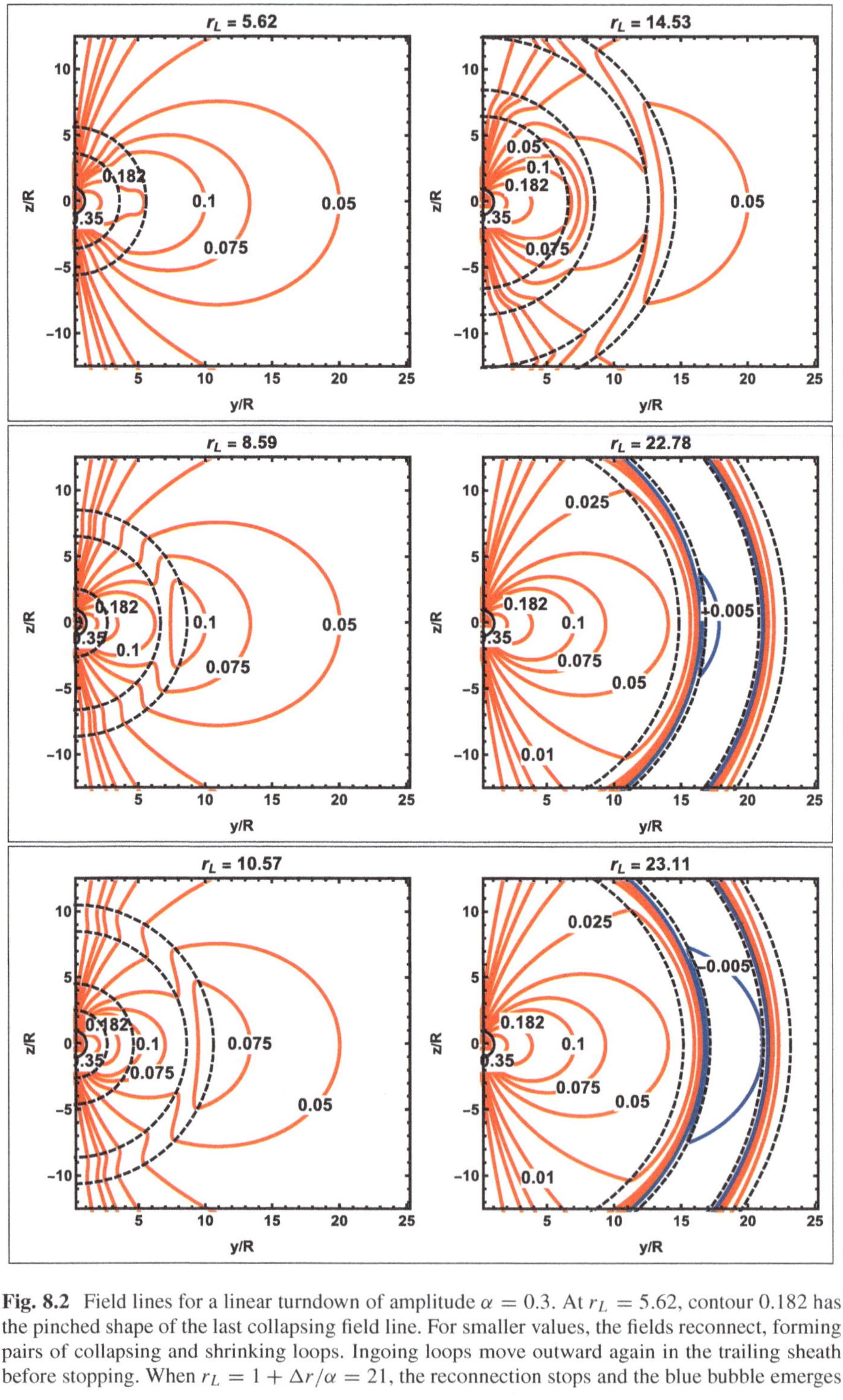

Fig. 8.2 Field lines for a linear turndown of amplitude $\alpha = 0.3$. At $r_L = 5.62$, contour 0.182 has the pinched shape of the last collapsing field line. For smaller values, the fields reconnect, forming pairs of collapsing and shrinking loops. Ingoing loops move outward again in the trailing sheath before stopping. When $r_L = 1 + \Delta r/\alpha = 21$, the reconnection stops and the blue bubble emerges

carried outward in the trailing sheath. By $r_L = 22.78$, all of the original dipole field lines have been pinched off and the blue, reversed-polarity field is becoming visible at the front of the trailing sheath and expanding outward into the middle region between the sheaths. At later times, these blue contours will continue to emerge and eventually indicate the outgoing bubble as it leaves the region.

8.2 A Plane Wave Moving Past a Dipole Field

Next, we examine the field of a plane wave moving past a dipole field in two dimensions. For simplicity, we begin using the (r, ϕ, z) cylindrical coordinate system and consider a vector potential of the form $\mathbf{A} = A_z(r, \phi)\ \mathbf{e_z}$. In this case, the magnetic field is given by

$$\mathbf{B} = \boldsymbol{\nabla}\mathbf{x}\mathbf{A} = \frac{1}{r}\frac{\partial A_z}{\partial \phi}\ \mathbf{e_r} - \frac{\partial A_z}{\partial r}\ \mathbf{e_\phi}. \tag{8.4}$$

For this static field, $\boldsymbol{\nabla}\mathbf{x}\mathbf{B} = \mathbf{0}$ everywhere except at the location of the source of the field, which we take to be a circle of unit radius. Consequently,

$$\nabla^2 A_z = \frac{1}{r}\frac{\partial}{\partial r}\left(r\frac{\partial A_z}{\partial r}\right) + \frac{1}{r^2}\frac{\partial^2 A_z}{\partial \phi^2} = 0. \tag{8.5}$$

By separating variables, we obtain a solution of the form

$$A_z(r, \phi) = r^m\ e^{im\phi}, \tag{8.6}$$

where m is a constant of integration. For simplicity, we choose the solution:

$$A_z(r, \phi) = \begin{cases} -(\cos\phi)/r, & r \geq 1 \\ -r\ \cos\phi, & r \leq 1, \end{cases} \tag{8.7}$$

where r is expressed in units of the source-circle radius and A_z is expressed in units of the constant value A_0. This vector potential gives rise to the magnetic field

$$\mathbf{B}(r, \phi) = \begin{cases} (\sin\phi)/r^2\ \mathbf{e_r} - (\cos\phi)/r^2\ \mathbf{e_\phi}, & r \geq 1 \\ \sin\phi\ \mathbf{e_r} + \cos\phi\ \mathbf{e_\phi}, & r \leq 1, \end{cases} \tag{8.8}$$

where the field $\mathbf{B}(r, \phi)$ is expressed in units of $B_0 = A_0/R$ and R is radius of the circle—the scale in which all distances r are measured. As expected for a field with $\boldsymbol{\nabla}\cdot\mathbf{B} = 0$, the radial component of $\mathbf{B}$ is continuous at $r = 1$. Also, we can see that the azimuthal component is not continuous, but has a jump of $-2\cos\phi$. This must equal the current normal to the plane, which is into the plane when $\cos\phi > 0$ and out of the plane when $\cos\phi < 0$ (i.e., when $x > 0$ and $x < 0$, respectively).

To derive the field lines outside $r = 1$, we write

$$\frac{dr}{r\,d\phi} = \frac{B_r}{B_\phi} = -\frac{\sin\phi}{\cos\phi}, \tag{8.9}$$

whose solution is $r = r_0 \cos\phi$, where r_0 is a constant of integration. The reader may recognize this solution as being similar to the equation for the field lines of a dipole in three dimensions, which is $r = r_0 \sin^2\theta$. If θ were measured from the equator rather than the pole, then $\sin\theta$ would become $\cos\theta$ and the two formulas would differ only in the power of the cosine. The single power of $\cos\phi$ causes the field lines to be circles rather than the elongated loops of the dipole field in three dimensions that occur for $\cos^2\phi$. To see this, we transform to the xyz rectangular coordinate system for which $x = r \cos\phi$ and $y = r \sin\phi$. In this case, the field-line equation becomes $r^2 = r_0\, x$. This is equivalent to $(x - r_0/2)^2 + y^2 = (r_0/2)^2$, which is the equation of a circle of radius $r_0/2$ centered at $(r_0/2, 0)$. Thus, although r_0 was defined to be the value of r when $\phi = 0$, it turns out to be the diameter of the circle and not the radius. Consequently, in the rest of our discussion, we shall replace r_0 by the symbol D_0 to represent the diameters of the circular field lines.

Next, we convert the field and vector potential to rectangular coordinates so that they can easily be combined with the field of a plane wave which we anticipate will be expressed in rectangular coordinates. The vector potential given by Eq. (8.7) becomes

$$A_z(x, y) = \begin{cases} -x/(x^2 + y^2), & x^2 + y^2 \geq 1 \\ -x, & x^2 + y^2 \leq 1, \end{cases} \tag{8.10}$$

which gives rise to the magnetic field components

$$B_x = (\sin 2\phi)/r^2 = 2xy/(x^2 + y^2)^4 \tag{8.11a}$$

$$B_y = -(\cos 2\phi)/r^2 = (y^2 - x^2)/(x^2 + y^2)^4 \tag{8.11b}$$

outside the circle of unit radius. Inside the circle, the components are

$$B_x = 0 \tag{8.12a}$$

$$B_y = 1, \tag{8.12b}$$

corresponding to a uniform field directed upward along the y-axis. Thus, the field lines extend upward through the unit circle and then bend around in circular arcs of radius $D_0/2$ before returning to the unit circle.

We can also derive the field lines from $A_z(x, y)$. If we hold $A_z(x, y)$ constant while varying the values of x and y, we have the relation

$$dA_z(x, y) = \frac{\partial A_z}{\partial x}\,dx + \frac{\partial A_z}{\partial y}\,dy = 0. \tag{8.13}$$

Then, we can use the equation

$$\mathbf{B} = \nabla \mathbf{x} \mathbf{A} = \frac{\partial A_z}{\partial y} \mathbf{e_x} - \frac{\partial A_z}{\partial x} \mathbf{e_y} \tag{8.14}$$

to express the partial derivatives of the vector potential in terms of the magnetic field components, B_x and B_y. Substituting these expressions into Eq. (8.13), we obtain

$$dA_z(x, y) = -B_y\, dx + B_x\, dy = 0, \tag{8.15}$$

which is another form of the field-line equation

$$dy/dx = B_y/B_x. \tag{8.16}$$

Thus, the contours of constant $A_z(x, y)$ are magnetic field lines.

At this point, it is instructive to introduce the magnetic "flux," Φ, defined by

$$\Phi(x, y) = \int_x^\infty B_y(x', y)\, dx'. \tag{8.17}$$

This magnetic flux is expressed in units of $B_0 R = A_0$. Substituting $B_y(x', y) = -\partial A_z(x', y)/\partial x'$ and performing the integration, we obtain

$$\Phi(x, y) = A_z(x, y) - A_z(\infty, y) = A_z(x, y). \tag{8.18}$$

Referring to Eq. (8.10), we see that the flux $\Phi(1, 0) = -1$, corresponding to all of the field lines that cross the equator in the $-y$ direction outside the circle when $x > 1$. An equal amount of positive flux lies inside the circle between the origin and $x = 1$, as one can deduce from $\Phi(0, 0) = 0$. Likewise because $\Phi(-1, 0) = 1$, the same amount of flux lies inside the circle between $x = -1$ and the center. This unit of flux is $B_0 R = B_0 = 1$. Thus, unlike the unit of flux that we used in describing the spherical dipole field, for which 1 unit occurred inside the sphere, 2 units of flux lie inside the circle for this quasi-dipole field in the xy-plane.

The contours of constant flux are easily deduced from Eqs. (8.10) and (8.18). In this case, the field-line equation becomes $\{x - 1/(2\Phi)\}^2 + y^2 = \{1/(2\Phi)\}^2$ outside the sphere. This equation describes a circle of radius $1/(2|\Phi|)$ centered at $(1/(2\Phi), 0)$. Thus, outside the unit circle, the contours of flux, Φ, are circles of diameter $D_0 = 1/|\Phi|$, and inside they are vertical lines at $x = -\Phi$.

Next, we look for a plane wave that we can combine with this "dipole" field. For this purpose, we consider a vector potential, $\mathbf{A}^w$, whose only component is along the z-axis and is given by

$$A_z^w(x, x_L) = a\, f(x - x_L), \tag{8.19}$$

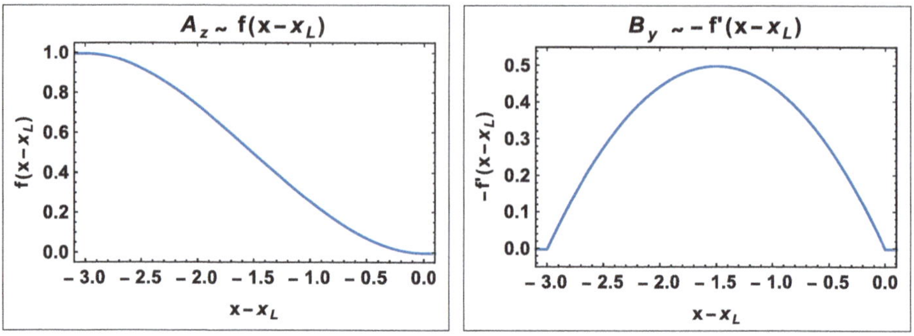

Fig. 8.3 Profiles of the wave field, showing its vector potential, A_z (left), normal to the xy-plane, and its vertical component, B_y (right), in the xy-plane. Whereas A_z increases gradually from 0 to 1 across the 3-unit width of the wave, B_y reaches a maximum value of $+0.5$ when $x - x_L = -1.5$ at the center of the wave

where a is a constant and x_L is time indicated by the position of the leading edge of the wave (analogous to r_L for the spherical wave). The function $f(x - x_L)$ satisfies the wave equation $\partial^2 f/\partial x^2 - \partial^2 f/\partial x_L^2 = 0$, and is defined below. For this vector potential, the wave field has a single component given by

$$B_y^w = -\partial A_z^w/\partial x = -a\ f'(x - x_L). \tag{8.20}$$

For simplicity, we select the function, f, to be a cubic polynomial given by

$$f(x - x_L) = \begin{cases} 0, & x - x_L \geq 0 \\ 2\left(\frac{x - x_L}{3}\right)^3 + 3\left(\frac{x - x_L}{3}\right)^2, & 0 > x - x_L > -3 \\ 1, & x - x_L \leq -3. \end{cases} \tag{8.21}$$

As shown in the left panel of Fig. 8.3, this function increases smoothly from 0 at $x = x_L$ to 1 at $x = x_L - 3$. Also, as shown in the right panel, the quantity $-f'(x - x_L)$, which is proportional to B_y, is positive and reaches its peak value of 0.5 at the center of this 3-unit interval where $x - x_L = -1.5$. With these definitions, we have created a plane wave whose magnetic field is oriented in the $+y$ direction and is confined to a 3-unit sheath that moves in the $+x$ direction at the speed of light. The peak amplitude of this wave is $0.5a$, where a is a constant that we shall set equal to 0.15 in the remainder of this chapter.

Next, because the field equations are linear, we can add the "dipole" field given by Eqs. (8.11) and (8.12) to the field of the plane wave in Eq. (8.20) to obtain the combined field. The result is

$$\mathbf{B} = \begin{Bmatrix} \left[2xy/(x^2 + y^2)^2\right] \mathbf{e_x} + \left[(y^2 - x^2)/(x^2 + y^2)^2\right] \mathbf{e_y}, & x^2 + y^2 \geq 1 \\ \mathbf{e_y}, & x^2 + y^2 \leq 1 \end{Bmatrix}$$
$$-a\ f'(x - x_L)\ \mathbf{e_y}. \tag{8.22}$$

Likewise the individual vector potentials can be added to obtain the vector potential for the combined field.

$$A_z(x, y, x_L) = \left\{ \begin{array}{cc} -x/(x^2+y^2), & x^2+y^2 \geq 1 \\ -x, & x^2+y^2 \leq 1 \end{array} \right\} + a\ f(x - x_L). \quad (8.23)$$

For this combined vector potential, $A_z(\infty, y, x_L)$ vanishes, so that $\Phi(x, y, x_L) = A_z(x, y, x_L)$. Thus, the contours of constant vector potential are still contours of constant flux.

However, because these contours change with time, the question arises of whether they are still magnetic field lines. In particular, with this time dependence, the change of vector potential is

$$dA_z(x, y, x_L) = \frac{\partial A_z}{\partial x}\,dx + \frac{\partial A_z}{\partial y}\,dy + \frac{\partial A_z}{\partial x_L}\,dx_L, \quad (8.24)$$

which becomes

$$dA_z(x, y, x_L) = -B_y\,dx + B_x\,dy + \frac{\partial A_z}{\partial x_L}\,dx_L \quad (8.25)$$

when the magnetic field components, $B_x = \partial A_z/\partial y$ and $B_y = -\partial A_z/\partial x$, are inserted. Thus, in Eq. (8.25), the extra term, $(\partial A_z/\partial x_L)\,dx_L$, prevents $dA_z(x, y, x_L)$ and $-B_y\,dx + B_x\,dx$ from vanishing together. It is interesting to recall that this change of vector potential corresponds to an electric field $E_z = -\partial A_z/\partial x_L$ (with E_z in units of $E_0 = A_0c/R = cB_0$). Consequently, we can think of the extra term in Eq. (8.25) as being due to the presence of this electric field. Then we can blame this electric field for the apparent discrepancy between contours of magnetic flux and magnetic field lines.

To see the effect of this term, we begin by keeping x_L fixed at a given time, say x_{L1}, prior to the arrival of the wave. In this case, $dx_L = 0$, and the extra term vanishes, making the flux contours and magnetic field lines the same. We select a particular flux, Φ_1, and let the coordinates (x, y) range over their possible values, to produce the contour $\Phi(x, y, x_{L1}) = \Phi_1$. As mentioned above for the field given by Eqs. (8.10)–(8.12), this contour corresponds to a circular field line of diameter $D_1 = 1/\Phi_1$ outside the unit circle and a vertical line defined by $x = -\Phi_1$ inside the unit circle.

Next, we let time advance to the value $x_L = x_{L2}$ after the passage of the wave, and then ask where the contour labeled Φ_1 is and what contour moved into its previous position. The contour of flux, Φ_1, becomes $\Phi(x, y, x_{L2}) = \Phi_1$, which is different from the original contour, but identical to the field line given by $\Phi(x, y, x_{L1}) = \Phi_1 - a$, where a is the change of A_z during the passage of the wave. These contours are the same because $d\Phi = \Phi(x, y, x_{L2}) - \Phi(x, y, x_{L1}) = \Phi_1 - (\Phi_1 - a) = a$, which means that $-B_y\,dx - B_x\,dy = 0$. Thus, as the wave passed the contour labeled Φ_1, that contour moved to the position of the contour that was labeled $\Phi_1 - a$ at time x_{L1}. Consequently, its new diameter is $D_2 = 1/(\Phi_1 - a)$.

Likewise, the location of the vertical line inside the unit circle moved from $x = -\Phi$ to $x = -(\Phi_1 - a)$. This means that when $\Phi > a$, the contours for $x < 0$ increase their diameters and shift their vertical lines closer to the origin.

Having seen where the contour labeled Φ_1 goes, we next consider which of the initial contours moved in to take its place. To see this, we recognize that $\Phi(x, y, x_{L1}) = \Phi_1$ and $\Phi(x, y, x_{L2}) = \Phi_1 + a$ are the same curves. This is a consequence of $dA_z = \Phi(x, y, x_{L2}) - \Phi(x, y, x_{L1}) = (\Phi_1 + a) - \Phi_1 = a$, and therefore $-B_y\, dx + B_x\, dy = 0$. Thus, after the passage of the wave, the contour labeled Φ_1 at time x_{L1} is replaced by the contour that was initially labeled $\Phi_1 + a$.

Figure 8.4 shows the flux contours as the plane wave (defined by the vertical dashed lines) moves inward toward the center. The red and blue lines are contours of positive and negative flux, respectively. They do not indicate the direction of the field, which points upward in the $+y$ direction inside the circle and bends downward at the "equator" outside the circle, like the field lines of a dipole field of positive polarity.

As shown in the right panel of Fig. 8.3, the field of the plane wave is peaked at the center of the 3-unit interval and vanishes at the ends of the interval. It is a field of positive polarity pointing in the $+y$ direction. Consequently, as this wave of positive polarity moves inward and encounters the weak negative field at the outer fringes of the dipole field, neutral points form near the front and back of the plane wave. The accompanying field line reconnection opens the closed loops at the front of the wave, like the ones marked 0.15 in the first panel of Fig. 8.4. The reconnection at the back of the wave creates new closed loops, like the one marked 0.29 in the third panel. As explained above, the contour marked 0.15 moves to the location of the current contour marked 0.15–0.15 = 0 and lies directly over the pole.

As the wave moves further inward and encounters regions of stronger negative field, the neutral points move closer to the center of the wave where its positive field is correspondingly strong. By $x_L = -1.7$, the peak field strength has been reached, and a neutral point can no longer form. At this time, the contour marked 0.37 is stretching outward from its initial position at $D_1 = 1/0.37 = 2.70$ to replace the contour that was originally marked 0.37–0.15 = 0.22 and had the diameter $D_2 = 1/0.22 = 4.54$.

This motion toward the advancing wave is what typically happens when a field of one polarity penetrates a region of opposite polarity. We have seen it in the leading sheaths of the transients produced by current off-ramps and at the trailing sheaths of the transients produced by on-ramps. Likewise, we see it again in Fig. 8.5, which is a continuation of the time sequence in Fig. 8.4. As the wave moves into the region of positive x, the contours move inward in the opposite direction. By $x_L = 4.75$, the contour marked -0.22 has the characteristic stretched shape of the last loop to move inward without pinching off first. Its final diameter is $D_2 = 1/0.37 = 2.70$, corresponding to the position of the loop that was originally labeled –0.22–0.15 = -0.37.

As the wave moves outward, the contours marked with fluxes in the range 0.0–0.15 are carried into the region of positive x. They start to reconnect toward the end of the sequence, producing a black contour of 0 flux at the location of the former

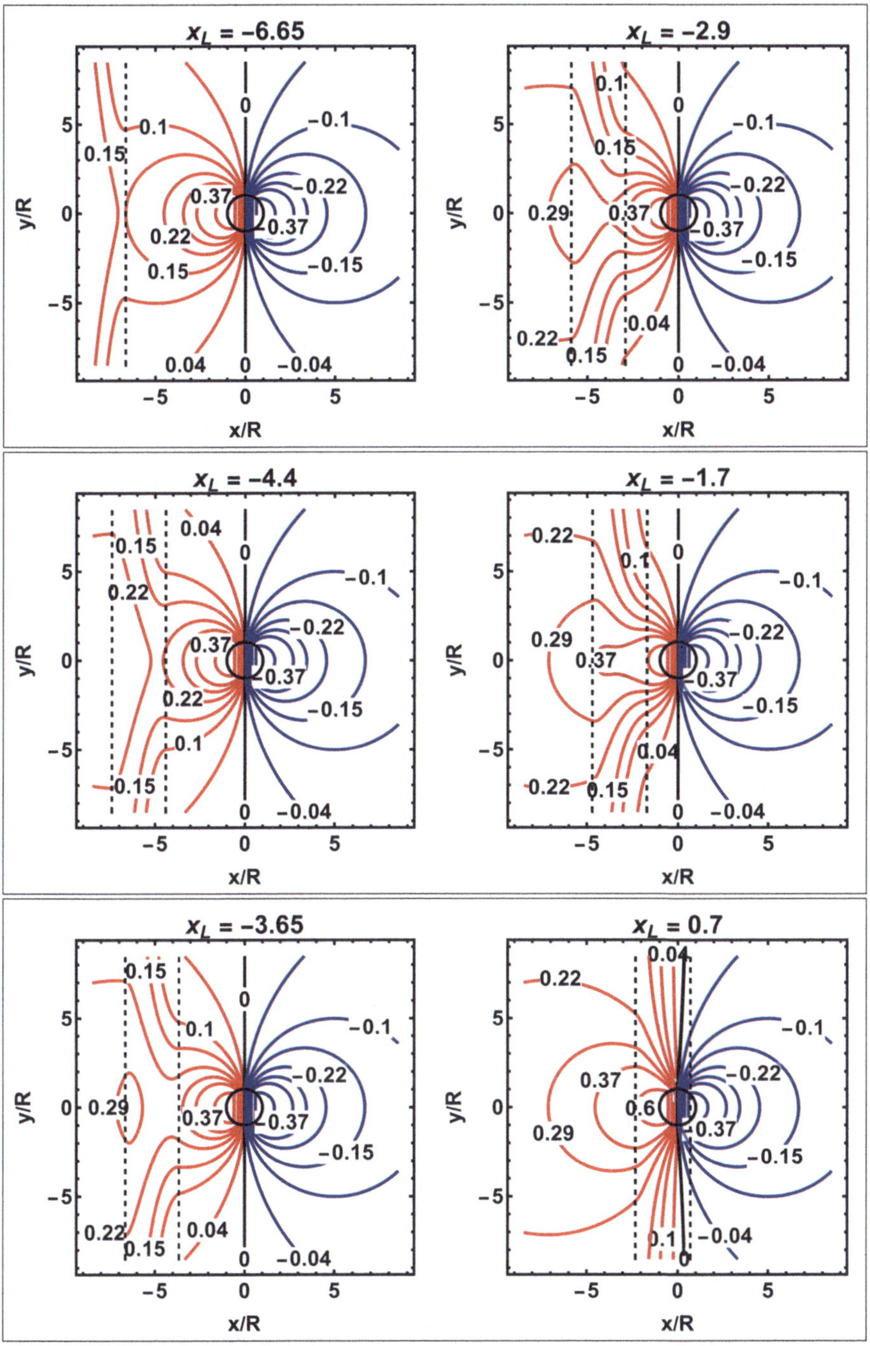

Fig. 8.4 The evolution of magnetic flux contours produced by a plane wave of y-directed field moving rightward through a quasi-dipole field of positive polarity. The red and blue colors refer to the positive and negative signs of the flux contours, not to the directions of the field lines

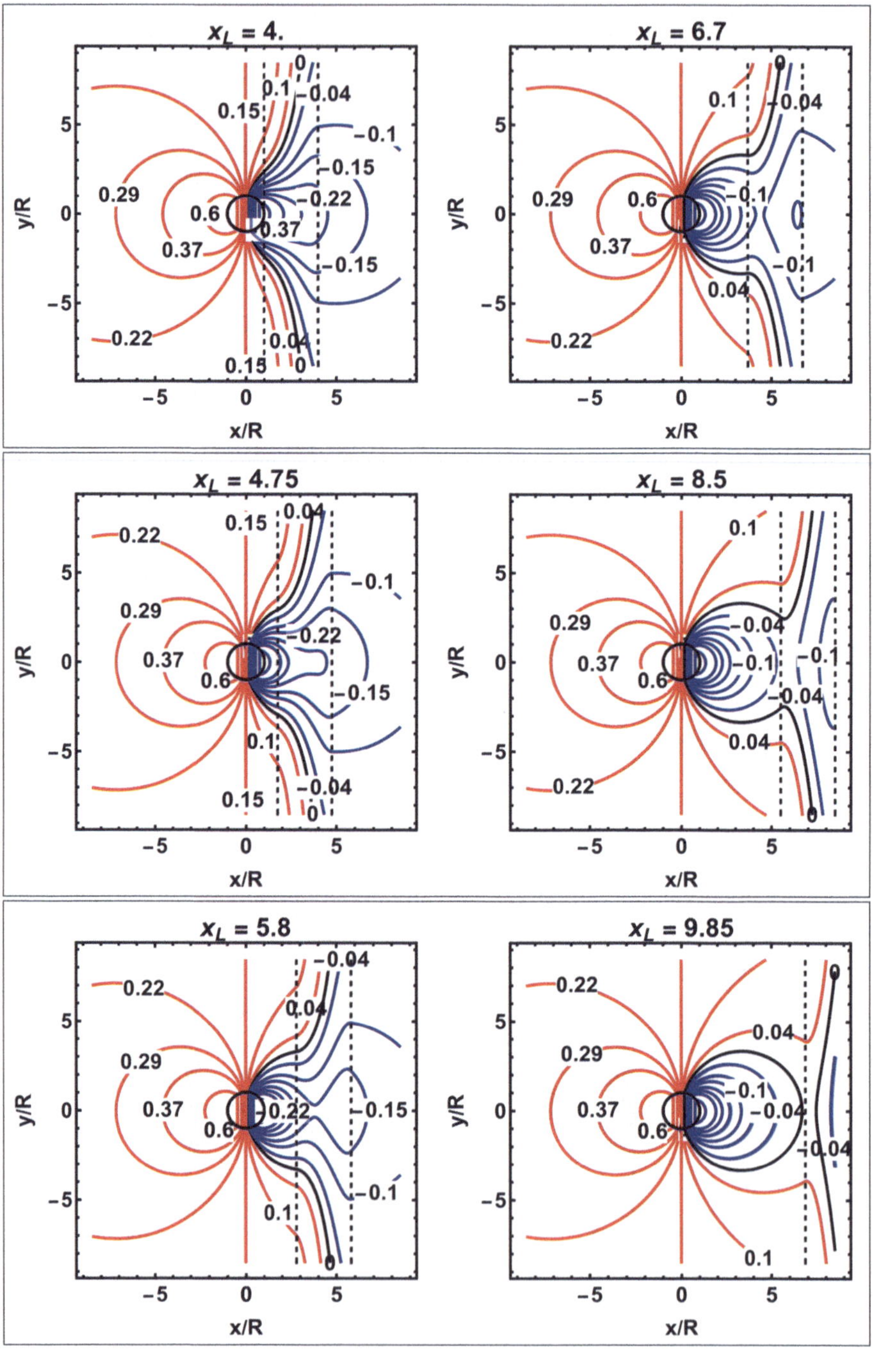

Fig. 8.5 A continuation of the sequence in Fig. 8.4, showing the magnetic flux contours as the plane wave moved outward from the center of the dipole field

field line marked −0.15 and a quasi-vertical contour in the outgoing wave. Thus, the original black contour marked 0 and extending directly upward over the pole in Fig. 8.4 became circular and moved to the location of this former field line of diameter $D_2 = 1/0.15 = 6.67$.

The net result is that the contours in the negative sector have spread out and the contours marked 0–0.15 have moved into the positive sector, causing the negative contours to crowd closer together. This is somewhat analogous to the transfer of flux from the nose of Earth's magnetosphere to its tail when the solar wind carries a southward-directed field past the Earth. However, in that case, the Earth's rotation carries the downwind field lines back to the upwind side and restores the symmetry.

Next, we derive the contours of constant flux along the x-axis. We obtain these contours by setting $y = 0$ in Eq. (8.23)

$$\Phi(x, 0, x_L) = \left\{ \begin{array}{ll} -1/x, & x^2 \geq 1 \\ -x, & x^2 \leq 1 \end{array} \right\} + a\, f(x - x_L). \tag{8.26}$$

Proceeding as we did for other space-time maps, we obtain the neutral point curves by setting $\partial\Phi/\partial x = 0$, and then we find the starting locations by combining this equation with $\partial^2\Phi/\partial x^2 = 0$. Because we expect these starting points to lie near the center of the wave, we define the offset distance δ by $x = x_L - 1.5 + \delta$ and write these two equations in terms of the variables x and δ:

$$1/x^2 + 2a\{(\delta/3)^2 - 1/4\} = 0 \tag{8.27a}$$

$$-1/x^3 + (2a/3)(\delta/3) = 0. \tag{8.27b}$$

Proceeding as before, we obtain an approximate solution expressed as a power series in $\gamma = a/2$:

$$\delta = \frac{9}{4}\gamma^{1/2} - \frac{243}{32}\gamma^{3/2} + \frac{28431}{512}\gamma^{5/2}, \tag{8.28a}$$

$$x = \gamma^{-1/2} + \frac{9}{8}\gamma^{1/2} - \frac{729}{128}\gamma^{3/2}, \tag{8.28b}$$

$$x_L = 1.5 + \gamma^{-1/2} - \frac{9}{8}\gamma^{1/2}, \tag{8.28c}$$

$$\Phi = -\gamma^{1/2} + \gamma - \frac{9}{8}\gamma^{3/2}. \tag{8.28d}$$

Referring to Fig. 8.6, this is the solution in the upper part of the space-time map where $x > 0$. For $a = 0.15$, the equations give $\delta_1 = 0.54$, $x_1 = 3.84$, $x_{L1} = 4.88$, and $\Phi = -0.22$. The solution in the lower part of the map where $x < 0$ is easily obtained by setting $\delta_2 = -\delta_1$, $x_2 = -x_1$, $x_{L2} = 3 - x_{L1}$, and $\Phi_2 = a - \Phi_1 = 2\gamma - \Phi_1$. These two points lie on opposite sides of the center line, $x = x_L - 1.5$, and on flux contours whose magnitudes differ by $a = 0.15$.

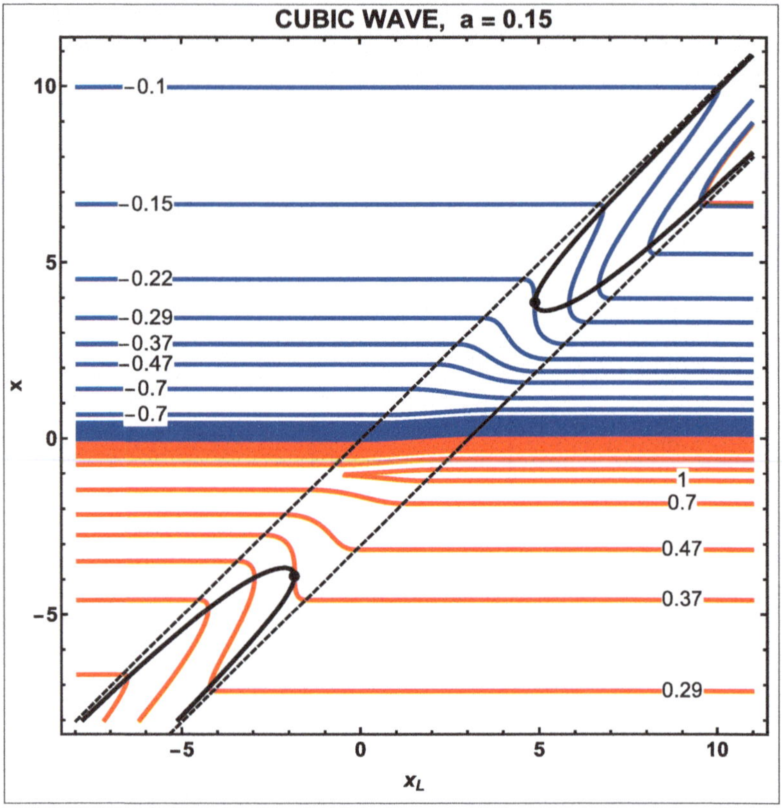

Fig. 8.6 A space-time map of magnetic flux contours for the field shown in Figs. 8.4 and 8.5, as they move along the x-axis. As in the previous figures, the red and blue tracks are contours of positive and negative flux, respectively. The dotted lines indicate the boundaries of the wave, the solid black curves are the *loci* of neutral points, and the black dots indicate where those neutral points are born

Figure 8.6 captures the motions of the flux contours along the x-axis. The two black, U-shaped curves show the formation of neutral points, initially along the boundaries of the inbound wave, and finally near the middle where the field is strong. This process occurs in reverse for the outbound wave where the neutral points are born near the middle and then separate as the wave moves outward into the weaker fringes of the dipole field. Between the two black curves, the slopes are negative (corresponding to motion opposite to the direction of the wave), except inside the unit circle. In this region, the fields of the wave and the dipole are parallel and the slopes are very slightly positive, corresponding to a small drift in the direction of the wave.

As in our other examples, perhaps the most interesting behavior is shown in the regions of the two black U-shaped curves where the neutral points occur. In these regions, the tracks have switchbacks, corresponding to apparent motions backwards in time. As we have seen, these switchbacks are a consequence of the reconnecting topology of the field, in which two new contours are produced near the trailing edge of the wave. One of these tracks continues on to become the loop left after the wave. The other track moves in the positive x-direction to meet and annihilate the inbound component at the leading edge of the wave.

The overall effect of these motions is to break the symmetry of the horizontal tracks that existed across the $x = 0$ line before the arrival of the wave. After the passage of the wave, the positive (red) contours spread out, while the negative (blue) contours crowd closer together. These changes are accomplished by adding $a = 0.15$ to the initial flux value. For example, the $+0.37$ contour ends up at the x value of the original (unlabeled) $+0.22$ contour. Likewise, the -0.22 contour ends up at the initial position of the -0.37 contour. The 0 contour, indicated by contiguous red and blue curves in the upper right part of Fig. 8.6, replaces the original -0.15 contour. This heralds the arrival of the remaining positive (red) contours with fluxes in the range 0–0.15 that occur at greater values of x outside the field of view. For example, after the passage of the wave, a red contour marked $+0.05$ would occur at $x = 10$ where the blue contour marked -0.1 occurred prior to the arrival of the wave.

Thus, for a plane wave moving past a dipole field, the vector potential (and therefore the flux) increased by a constant amount, a, corresponding to the change of $f(x - x_L)$ from 0 to 1 in Eq. (8.20). The contours retained their circular shapes, but the additive constant in the vector potential caused the contours to move to positions that differed from their original locations prior to the passage of the wave. This resulted in a new set of field lines, still circles, but with different diameters and locations than expected for their labeled fluxes. In Figs. 8.4 and 8.5, this caused the positive contours to increase their diameters. At the same time, a units of positive flux were transferred across the y-axis into the "outbound" hemisphere, causing the negative contours to become smaller and more crowded.

8.3 Remaining Issues

In this chapter, we have considered two more examples of one magnetic field passing through another. Again, we have seen that if the fields have opposite signs, the net field weakens and the field lines move backwards toward the incoming field. This creates a neutral point where the field lines pinch together and change their topology. If the fields have the same sign, a neutral point is not created and the net field moves forward with the incoming field.

In retrospect, it seems that this would happen for any two vector fields that pass through each other, regardless of whether they are rigorously defined magnetic fields or arbitrary mathematical abstractions. The only requirement is for the strength of one field to decrease with distance faster than the strength of the other field, as happens, for example, when a wave field with a $1/r$ dependence moves outward through a dipole field with a $1/r^3$ dependence. However, if the approaching fields are magnetic fields and are prevented from interpenetrating by the presence of conducting material, then the fields would have to expel the conducting material or diffuse through it before they form a neutral point.

Our analysis revealed a problem with the field-line concept when the magnetic field changes with time and an electric field is therefore present. In Sect. 8.2, we found that contours of constant flux were still magnetic field lines when the field changed with time, but that it was necessary to relabel the flux contours after the change. It was as if the dipole field was altered by the passage of the wave, which should not happen in a vacuum. Although we could restore the correct contours by adding a second plane wave of opposite polarity immediately after the first wave, there would still be some doubt about the detailed changes as the waves pass the dipole field. In Chap. 13, we shall encounter a fully three-dimensional model, in which a bubble of flux from one sphere passes the dipole field of a second sphere without producing such field-line inconsistencies.

Chapter 9
Short-Wave Radiation

9.1 The Linear Current Ramp with $\Delta r \leq 2$

In Chap. 5, we discussed the fields of a linear current ramp whose width was greater than $\Delta r = 2$. In that case, the ingoing component had already reached the center of the sphere, turned around, and moved back outward across the sphere before its trailing edge left the current. This produced a long wave train that was shifted behind the corresponding wave train from the outgoing component by the relatively small distance of $2R$. Now, in order to study the shortwave properties of the radiation, we need to extend our discussion to ramps whose widths are less than $\Delta r = 2$.

We refer to Fig. 9.1, which shows the regions that are produced for a ramp of width $\Delta r = 1$. The boundary of the sphere is indicated by the solid black curve, and the leading and trailing boundaries of the two wave trains are colored red and blue, respectively. Proceeding counterclockwise from the upper left panel, we see that the blue trailing edges leave the surface when $2 - (r_L - \Delta r) = 1$, which is $r_L = 2$ for this example with $\Delta r = 1$. This blue circle produces a $1o/mo$ boundary outside the sphere and a $1ai/1bi$ boundary inside. The inbound wave fronts cross inside the sphere at a location obtained by setting the radius, $r_L - 2$, of the red outgoing component equal to the radius, $2 - (r_L - \Delta r)$, of the ingoing component. The result is $r_L = 2 + \Delta r/2$, which in this case, is $r_L = 2.5$, midway between the second and third frames.

As described in Chap. 5, we express the F-values as integrals of the profiles obtained for the sudden turn-on, and separated these integrals into "active" contri-

Electronic Supplementary Material The online version of this chapter (https://doi.org/10.1007/978-3-030-40264-8_9) contains supplementary material, which is available to authorized users.

N. R. Sheeley, Jr., *Transient Magnetic Fields*,
https://doi.org/10.1007/978-3-030-40264-8_9

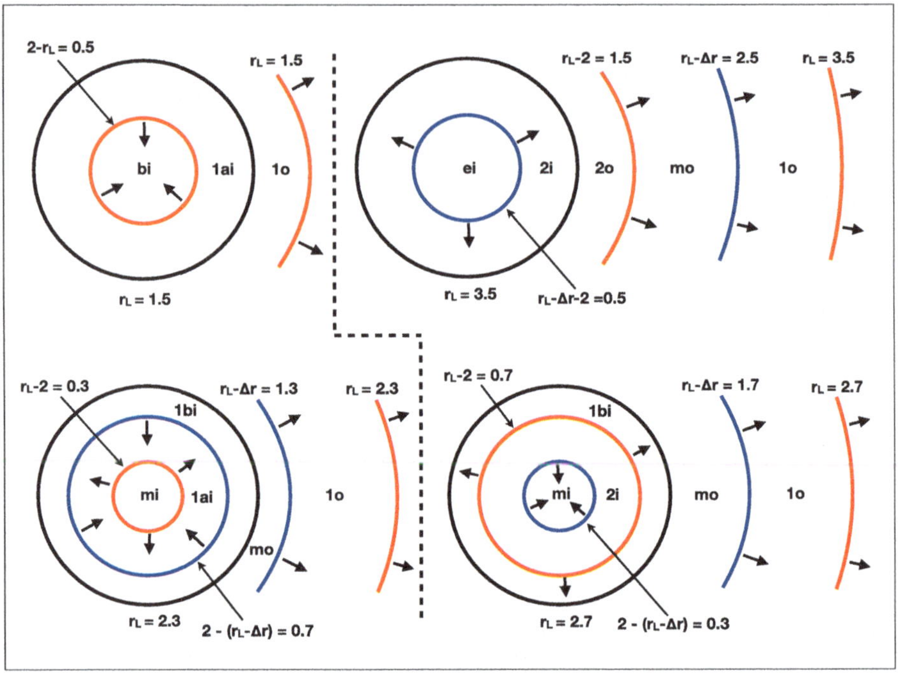

Fig. 9.1 A sketch of the regions produced by a linear current ramp of length $\Delta r = 1$ at times $r_L = 1.5,\ 2.3,\ 2.7,\ 3.5$, progressing counterclockwise from the upper left. The black circle represents the sphere of radius 1 in normalized units; the red and blue circles indicate the waves that are produced at the start and end of the ramp, respectively. Because $\Delta r < 2$, the blue wave enters the sphere before the red wave has left, causing the waves to pass through each other inside the sphere. This penetration replaces interior region $1ai$ with region $2i$ behind the red wave as it moves into remaining region $1bi$ and leaves the sphere

butions (either f_o or f_i) and "inactive" contributions (1 or r^3, respectively). Inside the sphere, we obtain:

$$
\begin{aligned}
F_{bi} &= 0, \\
F_{1ai} &= \tfrac{1}{\Delta r}\int_{2-r}^{r_L} f_i\big(r, r_L'\big)dr_L', \quad 2-(r_L-\Delta r) > 1 \\
F_{1bi} &= \tfrac{1}{\Delta r}\int_{r_L-\Delta r}^{r_L} f_i\big(r, r_L'\big)dr_L', \quad 2-(r_L-\Delta r) < 1 \\
F_{mi} &= \tfrac{1}{\Delta r}\int_{2-r}^{2+r} f_i\big(r, r_L'\big)dr_L' + \tfrac{1}{\Delta r}\int_{2+r}^{r_L} r^3 dr_L' = \left(\tfrac{r_L-1}{\Delta r}\right) r^3 \\
F_{2i} &= \tfrac{1}{\Delta r}\int_{r_L-\Delta r}^{2+r} f_i\big(r, r_L'\big)dr_L' + \tfrac{1}{\Delta r}\int_{2+r}^{r_L} r^3 dr_L' \\
F_{ei} &= \tfrac{1}{\Delta r}\int_{r_L-\Delta r}^{r_L} r^3 dr_L' = r^3,
\end{aligned}
\tag{9.1}
$$

where F_{bi} and F_{ei} are the beginning and ending values. Although the first of the $1i$ regions ($1ai$) forms as soon as the red wave enters the sphere at $r_L = 1$, the second $1i$ region ($1bi$) does not form until the blue wave enters the sphere at $2 - (r_L - \Delta r) = 1$.

Outside the sphere, we obtain:

$$\begin{aligned}
F_{bo} &= 0, \\
F_{1o} &= \tfrac{1}{\Delta r} \int_r^{r_L} f_o\big(r, r_L'\big) dr_L', \\
F_{mo} &= \tfrac{1}{\Delta r} \int_{r_L-\Delta r}^{r_L} f_o\big(r, r_L'\big) dr_L', \\
F_{2o} &= \tfrac{1}{\Delta r} \int_{r_L-\Delta r}^{r+2} f_o\big(r, r_L'\big) dr_L' + \tfrac{1}{\Delta r} \int_{r+2}^{r_L} dr_L', \\
F_{eo} &= \tfrac{1}{\Delta r} \int_{r_L-\Delta r}^{r_L} dr_L' = 1,
\end{aligned} \tag{9.2}$$

where F_{bo} and F_{eo} are the beginning and ending values.

It is easy to see that the expressions in Eqs. (9.1) and (9.2) reduce to those of Eqs. (5.6) and (5.7), respectively, when $\Delta r = 2$, and that $F(r, r_L)$ is continuous across the region boundaries. However, an even more powerful check on their validity can be obtained by using the U–V wave approach described in Chap. 5. In that approach, one can start with the f-values outside the sphere (which are easier to calculate), and use these values to derive the ones inside the sphere.

In Chap. 5, we regarded the individual boundaries as the leading edges of waves that add their own f-values to the regions lying ahead to produce the f-values in the regions behind them. Referring to Fig. 9.1, we define red-wave f-values, U, by examining the boundaries $bo/1o$ and $bi/1ai$:

$$U_a^{out}(r, r_L) = F_{1o} - F_{bo} = F_{1o}, \tag{9.3a}$$

$$U_b^{in}(r, r_L) = F_{1ai} - F_{bi} = F_{1ai}, \tag{9.3b}$$

$$U_b^{out}(r, r_L) = -F_{1ai}(-r, r_L). \tag{9.3c}$$

Likewise, we define the blue-wave f-values, V, from an examination of the boundaries $1o/mo$ and $2o/eo$ (which is not shown in Fig. 9.1 because it does not form until the second blue wave front, V_b^{out}, leaves the sphere):

$$V_a^{out}(r, r_L) = F_{mo} - F_{1o}, \tag{9.4a}$$

$$V_b^{out}(r, r_L) = F_{eo} - F_{2o} = 1 - F_{2o}. \tag{9.4b}$$

Next, examining the four boundaries $ei/2i$, $2o/mo$, $mi/1ai$, and $1bi/2i$, we can write

$$V_b^{out}(r, r_L) + F_{2i} = r^3, \tag{9.5a}$$

$$U_b^{out}(r, r_L) + F_{mo} = F_{2o}, \tag{9.5b}$$

$$U_b^{out}(r, r_L) + F_{1ai} = F_{mi}, \tag{9.5c}$$

$$U_b^{out}(r, r_L) + F_{1bi} = F_{2i}. \tag{9.5d}$$

Substituting the values of U_b^{out} and V_b^{out} from the previous sets of equations, we obtain expressions for the four interior f-values:

$$F_{2i} = F_{2o} + r^3 - 1, \tag{9.6a}$$

$$F_{1ai} = F_{mo}(-r, r_L) - F_{2o}(-r, r_L), \tag{9.6b}$$

$$F_{mi} = F_{1ai} - F_{1ai}(-r, r_L), \tag{9.6c}$$

$$F_{1bi} = F_{2i} + F_{1ai}(-r, r_L), \tag{9.6d}$$

where the F-values are evaluated at $(-r, r_L)$ when indicated. Otherwise, they are evaluated at (r, r_L). After some algebra, these equations reduce to

$$F_{2i} = F_{2o} + r^3 - 1, \tag{9.7a}$$

$$F_{1ai} = F_{mo}(-r, r_L) - F_{2o}(-r, r_L), \tag{9.7b}$$

$$F_{mi} = F_{1ai} - F_{mo} + F_{2o} = \left(\frac{r_L - 1}{\Delta r}\right) r^3, \tag{9.7c}$$

$$F_{1bi} = F_{mo} + r^3 - 1. \tag{9.7d}$$

These equations provide a relatively easy way to derive the f-values inside the sphere from the f-values obtained outside the sphere where there is no conceptually difficult "folding-back" process, and they give the relations listed in Eq. (9.1) for the f-values inside the sphere.

Next, we evaluate the integrals in Eqs. (9.1) and (9.2) using the expressions for f_i and f_o given by Eqs. (2.11a) and (2.11b). Inside the sphere, we obtain:

$$\begin{aligned}
F_{bi} &= 0, \\
F_{1ai} &= F_{1i} (\text{same as for } \Delta r \geq 2), \\
F_{1bi} &= f_i(r, r_L) + \left(\tfrac{\Delta r}{16}\right) [(\Delta r - 2\,(r_L - 1))^2 + 2\,(r_L - 1)^2 - 6\,(r^2 + 1)], \\
F_{mi} &= \left(\tfrac{r_L - 1}{\Delta r}\right) r^3, \\
F_{2i} &= F_{2i} \ (\text{same as for } \Delta r \geq 2), \\
F_{ei} &= r^3,
\end{aligned} \tag{9.8}$$

and outside, we find

$$
\begin{aligned}
F_{bo} &= 0,\\
F_{1o} &= F_{1o} \text{ (same as for } \Delta r \geq 2),\\
F_{mo} &= f_o(r, r_L) + \left(\tfrac{\Delta r}{16}\right) \left[(\Delta r - 2\,(r_L - 1))^2 + 2\,(r_L - 1)^2 - 6\,(r^2 + 1)\right],\\
F_{2o} &= F_{2o} \text{ (same as for } \Delta r \geq 2),\\
F_{eo} &= 1.
\end{aligned}
\tag{9.9}
$$

Note that the values of F_{1ai}, F_{2i}, F_{1o}, and F_{2o} in Eqs. (9.8) and (9.9) are identical to those listed in Eqs. (5.8) and (5.9) for the f-values when $\Delta r \geq 2$.

Looking back at Eq. (9.2), we can see that F_{mo} is the average value of f_o on the small interval $(r_L - \Delta r, r_L)$. Thus, it should be no surprise in Eq. (9.9) that F_{mo} is given by $f_o(r, r_L)$ plus some terms that depend on Δr. In fact, we can rewrite that equation with the extra terms included as a power series in Δr:

$$
F_{mo} = f_o(r, r_L) - \frac{3}{8}(\Delta r)\left\{r^2 + 1 - (r_L - 1)^2\right\} - \frac{1}{4}(\Delta r)^2\,(r_L - 1) + \frac{1}{16}(\Delta r)^3.
\tag{9.10}
$$

If Δr were so small that we could neglect it, then F_{mo} would be $f_o(r, r_L)$, which gives circulating field lines around an O-type neutral point whose radial position asymptotically approaches the location $r = r_L - 1$, midway through the interval $(r_L - 2, r_L)$. What if Δr is not that small? Suppose that it were necessary to include terms that are first order in Δr. In that case, the second term in Eq. (9.10) would have an effect.

If we think of averaging a collection of shifted circles over a small interval in a given direction, we would get another circle of approximately the same radius, but with its center shifted in that direction by one-half the length of the interval. Thus, it seems plausible that the second term in Eq. (9.10) will have the effect of shifting the asymptotic location of the O-type neutral point by an additional amount $\Delta r/2$. To test this idea, we evaluate $f_o(r, r_L - \Delta r/2)$:

$$
\begin{aligned}
f_o(r, r_L - \Delta r/2) = f_o(r, r_L) &- \frac{3}{8}(\Delta r)\left\{r^2 + 1 - (r_L - 1)^2\right\}\\
&- \frac{3}{16}(\Delta r)^2\,(r_L - 1) + \frac{1}{32}(\Delta r)^3.
\end{aligned}
\tag{9.11}
$$

Subtracting these two equations and rearranging the result, we obtain

$$
F_{mo} = f_o(r, r_L - \Delta r/2) - (r_L - 1)\left(\frac{\Delta r}{4}\right)^2 + 2\left(\frac{\Delta r}{4}\right)^3.
\tag{9.12}
$$

Thus, to first order in Δr, the asymptotic location of the O-type neutral point of the field generated from F_{mo} is located at the radial position $r = r_L - (\Delta r/2) - 1$, which is midway through the interval $(r_L - 2, r_L - \Delta r)$. A more precise calculation of the location of this O-type neutral point can be done by replacing r_L by $r_L - \Delta r/2$ in the expressions for x and r_0 as described in the paragraph between Eqs. (3.18) and (3.19).

Next, we use these f-values to plot contours of constant flux (field lines) when $\Delta r < 2$. Figure 9.2 compares a map obtained for $\Delta r = 0.5$ (left) with a corresponding map obtained for $\Delta r = 10$ (right) when both fields have evolved sufficiently to see their asymptotic features. For $\Delta r = 0.5$, a bubble of closed loops is centered between oppositely directed meridional fields in sheaths of width $\Delta r = 0.5$. An O-type neutral point lies near the center of the bubble where flux is removed. Also, an X-type neutral point occurs at the back of the trailing sheath where draped field lines reconnect, and offset the removal by adding flux to the bubble and to the dipole field left behind. For $\Delta r = 10$, the central region contains a growing dipole field bounded by oppositely directed meridional fields in sheaths of 2-unit width. In this case, the X-type neutral point also lies at the back of the trailing sheath, but the O-type neutral point lies at front of the sheath where the gradually rising loops are intercepted and swept away by the faster sheath.

It is instructive to interpret these field-line maps in terms of the wave trains produced by the current on-ramp. In the "short wave" case (i.e., $\Delta r = 0.5$), there are two wave trains of length $\Delta r = 0.5$ separated by a distance (or time in dimensionless units) $2 - \Delta r = 1.5$. One wave train always moves outward from the sphere. The other wave train starts at the same time as the first, but initially heads inward before turning around and moving outward 2 units behind the first. In Fig. 9.2, black dashed circles indicate the starting wave fronts of the two wave trains, and blue dashed circles indicate the ending wave fronts. Thus, we see a closely spaced pair of black and blue waves at the front of the outgoing disturbance and another closely spaced pair at the rear of the disturbance with a much larger region between them. However, this region between them is only $2 - \Delta r = 1.5$ units wide, which is comparable to the 2-unit transit time into and back out of the sphere. So this pattern is like the field of a sudden onset of current with a little bit of blurring at the start and end, like the field of the cubic on-ramp shown in the left panel of Fig. 7.13, but with a smaller spatial scale.

In the long-wave example (i.e., with $\Delta r = 10$), the wave trains are much longer. The initially ingoing wave train bends around at the center of the sphere and extends outward behind the initially outgoing wave train before the ends of either wave train have left the current source. Consequently, in the right panel of Fig. 9.2, we see a broad central region of width $\Delta r - 2 = 8$ units, lying between sheaths of 2-unit width. The leading sheath is bounded by a pair of black dashed circles corresponding to the starting wave fronts of the two trains, and the trailing sheath is bounded by a pair of blue dashed circles, corresponding to the ending wave fronts of these trains. As we have seen in Eq. (5.9), the f-value in the middle region is $F_{mo} = (r_L - 1)/\Delta r$, corresponding to a dipole field that increases linearly with time.

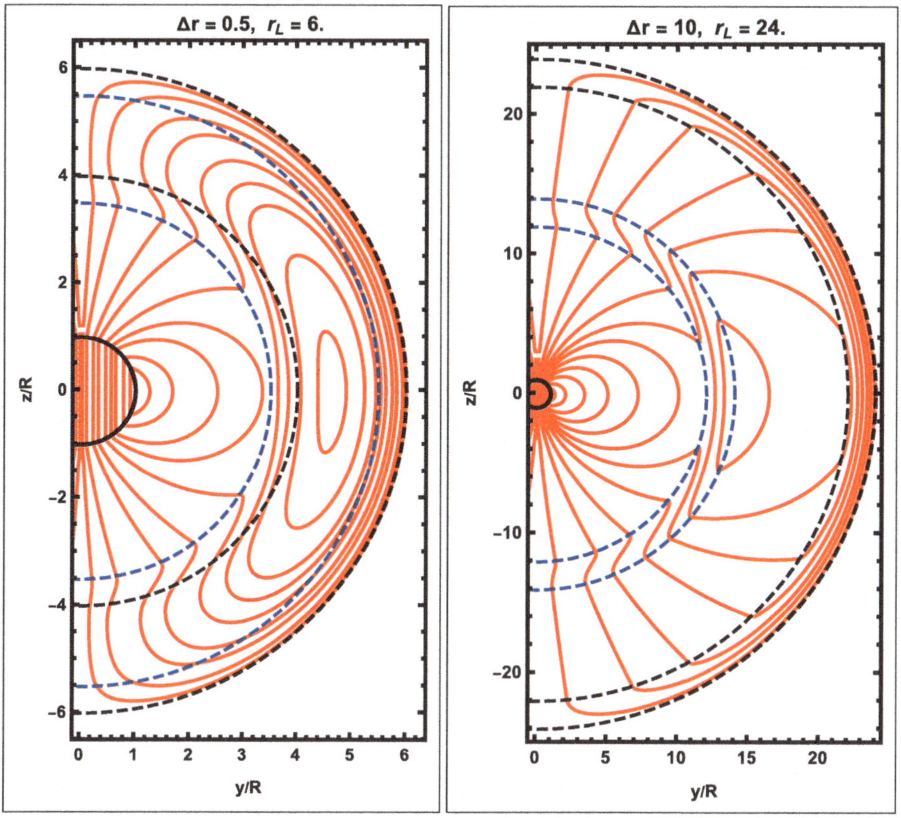

Fig. 9.2 Magnetic field lines for current on-ramps of width $\Delta r = 0.5$ (left) and $\Delta r = 10$ (right), comparing the topologies of the outgoing fields. For $\Delta r < 2$, a bubble of width $2 - \Delta r$ is sandwiched between two narrow sheaths of width Δr, but for $\Delta r > 2$, a growing dipole field of width $\Delta r - 2$ lies between sheaths of 2-unit width. Black dashed circles indicate the starting points of the outgoing and initially ingoing wave trains, and blue dashed circles indicate the ends of these wave trains. Behind them, a static dipole extends from a uniformly magnetized sphere of radius 1, whose location is indicated by a solid black circle

Figure 9.3 is a space-time map for the example with $\Delta r = 0.5$. The red lines are tracks of the contours of constant flux, showing the field-line motions along an equatorial strip where $\theta = \pi/2$. The black curve is the locus of neutral points where the red tracks have vertical slopes, and the black dot indicates the birth place of these neutral points. Perhaps as expected, this space-time map has features in common with the space-time maps for the symmetric cubic on-ramp (left panel of Fig. 7.12) and the space-time map for the sudden onset of current (Fig. 7.14). Like the symmetric cubic, the neutral-point curve starts near the $mo/2o$ boundary (*at* the boundary in this case) and its upper and lower branches extend to the center of the middle region (mo) and the back of the trailing region ($2o$), respectively. As we have seen in Fig. 9.2, this means that the X-type neutral point moves across the trailing

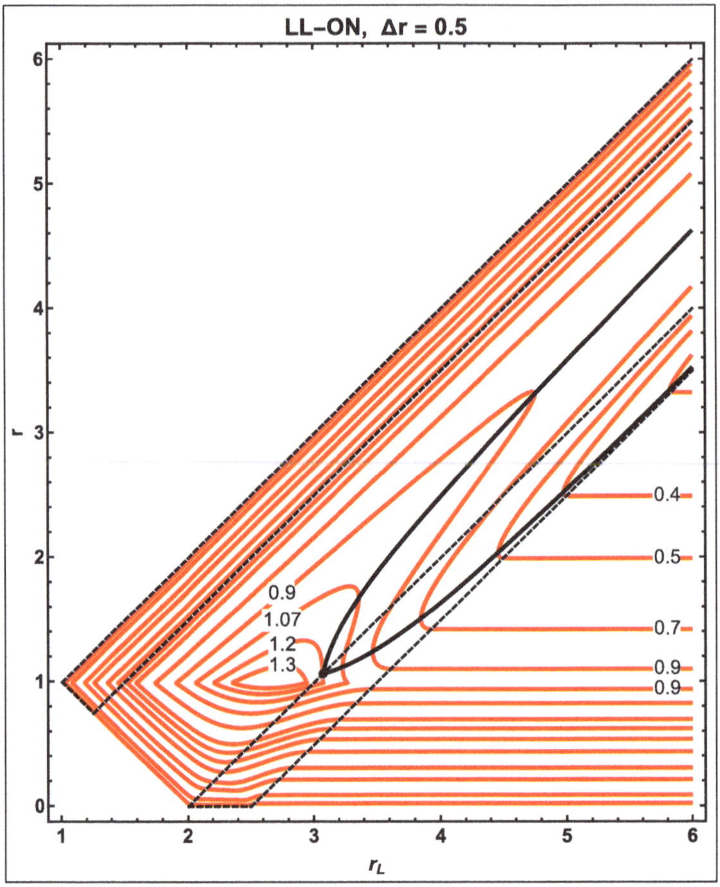

Fig. 9.3 Space-time tracks (red), obtained from contours of constant flux for a linear current on-ramp of width $\Delta r = 0.5$. These red curves indicate the motions of individual flux contours along the equator. The solid black curve indicates the *locus* of neutral points where the red tracks have vertical slopes, and the black dot indicates the starting point of this black curve, where the neutral points are first born. Black dashed lines mark the boundaries between the $1o$, mo, and $2o$ regions and their extensions inside the sphere

sheath to the rear of the disturbance and the O-type neutral point moves to the center.

However, for this short-wave on-ramp, the neutral points are born at $\Phi/\Phi_0 = 1.20$, which is close to the value of 1.23 obtained for the sudden turn-on, but far from the value of 0.39 obtained for the cubic on-ramp with $\Delta r = 6$. This means that the neutral points of this short-wave field are born close to the surface of the sphere where $r = 1$. Thus, when $\Delta r < 2$, and, in particular, when $\Delta r \to 0$, there is an "island" of closed tracks where the contours have values of $\Phi/\Phi_0 > 1$. In Fig. 9.3, these contours are labeled 1.3, 1.2, and 1.07. This excess flux must be removed from the sphere.

The value of $\Phi/\Phi_0 = 1.20$ marks the contour where the neutral points are first born, as indicated by the black dot in Fig. 9.3. For $\Phi/\Phi_0 > 1.2$, the external loops collapse and disappear at the surface where they meet their interior counterparts. However, for $\Phi/\Phi_0 \leq 1.2$ (but still > 1.0), the legs of the loops pinch together to produce an ingoing component, which quickly meets and annihilates its counterpart at the surface of the sphere, and an outgoing component, which forms the lower part of a closed loop that rapidly shrinks and disappears. Outside the sphere, where $\Phi/\Phi_0 < 1$, the ingoing component stops its inward motion as the trailing sheath passes by, and becomes a static loop of the sphere's exterior dipole field. The outgoing component gradually moves upward to become part of the shrinking loop that is nearly centered within the bubble. Unlike the field of the sudden turn-on, there is no region of Φ/Φ_0 where interior field lines simply "pop out" of the sphere and become part of a shrinking loop.

The space-time tracks for the sudden-onset field and the short-wave field are noticeably different along the region where the source switchbacks occur. Here, the slopes of the tracks are discontinuous for the sudden-onset field, but continuous for the short-wave field. Thus, for the sudden onset, the tracks do not become vertical at these locations, and the neutral-point curve does not have a lower branch. Nevertheless, as shown in Fig. 7.16, the draped field lines reconnect at the trailing edge of the disturbance, producing inward and outward components. Of course, in the limit that $\Delta r \to 0$, the properties of the short-wave field must reduce to those of the sudden-onset field. This is a consequence that we will use in the next section.

9.2 Radiated Flux and Energy for the Linear On-ramp

In Chap. 3 (Eq. (3.8)), we learned that the sudden current onset causes 3/4 of a unit of flux to be carried away in the outgoing bubble of closed loops. (The unit of flux $\Phi_0 = (\pi R^2)B_0$ is the amount of flux in the uniformly magnetized sphere of radius R and strength B_0.) Then, in Chap. 5 (Eq. (5.15)), we learned that the radiated flux is $1/\Delta r$ for a linear current ramp of width Δr, provided that $\Delta r \geq 2$. Now, we shall use the f-values in Sect. 9.1 to derive the radiated flux as a function of Δr when $\Delta r < 2$ and then combine these results into a single relation that is valid for all values of Δr.

Looking at the left panel of Fig. 9.2, we can see that the outgoing flux is spread over the entire region from $r_L - \Delta r - 2$ to r_L with field lines ahead of the O-type neutral point directed downward in the direction of increasing polar angle, θ, and field lines behind this neutral point directed upward. Also, we recognize that as r_L increases, the neutral point becomes centered between the leading and trailing sheaths. As $r_L \to \infty$, the reconnection behind the trailing sheath will continue to add flux to the trailing part of the bubble so that a balance asymptotically occurs between the leading and trailing parts. Consequently, for $\Delta r \leq 2$, we can expect this balanced flux to be given by

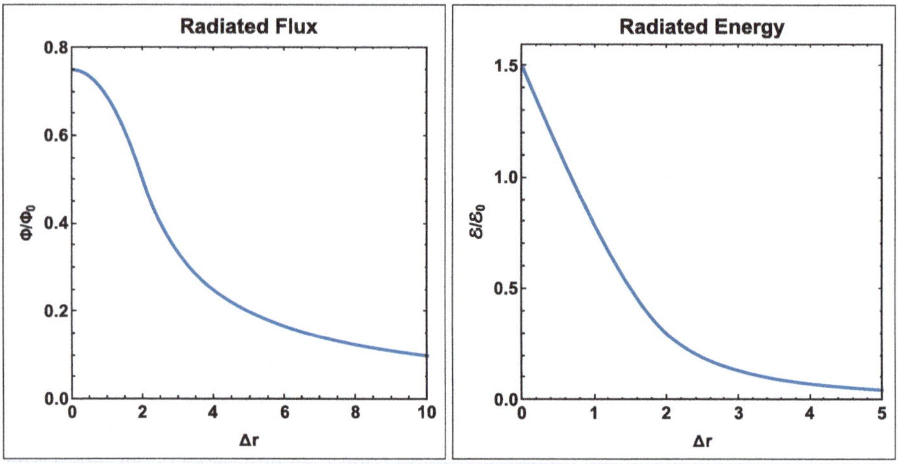

Fig. 9.4 Radiated flux (left) and energy (right) for a linear current on-ramp, plotted as a function of the ramp width, Δr. The flux and energy are expressed in terms of $\Phi_0 = (\pi R^2) B_0$ and $\mathcal{E}_0 = (\frac{4}{3}\pi R^3)(B_0^2/2\mu_0)$, the flux and energy inside the uniformly magnetized sphere of radius R and field strength B_0. The maximum radiated flux and energy are obtained for the sudden onset with $\Delta r = 0$

$$\Phi/\Phi_0 = \frac{F_{mo}(r_0, r_L, \Delta r)}{r_0} - \frac{F_{1o}(r_L, r_L)}{r_L} = \frac{F_{mo}(r_0, r_L, \Delta r)}{r_0}, \tag{9.13}$$

where $r_0 = r_L - 1 - \Delta r/2$. Taking F_{mo} from Eq. (9.9) and doing a little algebra, we find

$$\Phi/\Phi_0 = \frac{1}{2r_0} + \frac{3}{4} - \left(\frac{\Delta r}{4}\right)^2 \to \frac{3}{4} - \left(\frac{\Delta r}{4}\right)^2, \tag{9.14}$$

as $r_0 = r_L - 1 - \Delta r/2 \to \infty$.

As mentioned above, we used a similar procedure in Chap. 5 (see the discussion of Eq. (5.15)) to show that when $\Delta r \geq 2$ the amount of flux in each sheath approaches $1/\Delta r$ as $r_L \to \infty$. As expected, these short-wave and long-wave expressions for Φ/Φ_0 are equal at the boundary $\Delta r = 2$, and their derivatives are also equal. The left panel of Fig. 9.4 is a plot of the combined relation, showing the smooth decrease of radiated flux with the width, Δr, of the current ramp. Clearly, the more rapidly the current is turned on, the more flux is radiated away, up to the limit of 3/4 of a unit for the sudden onset.

Next, we consider the radiated energy. In Chap. 4, we derived an expression for the energy that flows outward across a spherical surface of radius r when the current was turned on suddenly. As indicated in Eq. (4.3), we obtained two terms—one representing the dipole energy that is ultimately stored outside the radius r, and the other representing the radiated energy. When the current is turned on gradually, a similar result is obtained, provided that the vector potential, f, for the sudden turn-on is replaced by, F, the vector potential for the current ramp, and that the limits of integration are changed to $r_{L1} = r$ and $r_{L2} = r + \Delta r + 2$, corresponding to the passage of the entire transient. Setting $F(r, r) = 0$ and $F(r, r + \Delta r + 2) = 1$, we obtain the dipole field energy, $1/2r^3$, and the radiated energy

$$\mathcal{E}_{rad}/\mathcal{E}_0 = -\frac{1}{r^2}\int_r^{r+\Delta r+2} \frac{\partial F}{\partial r}\frac{\partial F}{\partial r_L} dr_L, \tag{9.15}$$

where both terms are expressed in units of the energy, $\mathcal{E}_0 = (4\pi R^3/3)(B_0^2/2\mu_0)$, left in the uniformly magnetized sphere. As in the previous chapters, the distances r, Δr, and r_L are all expressed in units of the radius, R, of the sphere.

We can separate Eq. (9.15) into three parts corresponding to the contributions at the start ($1o$), the middle (mo), and the end ($2o$) of the transient. When $\Delta r \geq 2$, we obtain

$$\mathcal{E}_{rad}/\mathcal{E}_0 = -\frac{1}{r^2}\left[\int_r^{r+2} \frac{\partial F_{1o}}{\partial r}\frac{\partial F_{1o}}{\partial r_L} dr_L + \int_{r+2}^{r+\Delta r} \frac{\partial F_{mo}}{\partial r}\frac{\partial F_{mo}}{\partial r_L} dr_L \right.$$
$$\left. + \int_{r+\Delta r}^{r+\Delta r+2} \frac{\partial F_{2o}}{\partial r}\frac{\partial F_{2o}}{\partial r_L} dr_L \right]. \tag{9.16}$$

For a linear ramp, we may use the F-values given by Eq. (5.9). In this case, the middle region contributes no energy because $\partial F_{mo}/\partial r = 0$. The transition regions at the start and end of the ramp give $(1/\Delta r)^2\ (3/5 \pm 1/2r)$, where the $+$ and $-$ correspond to F_{1o} and F_{2o}, respectively. Thus, when $\Delta r \geq 2$, the total contribution from all three regions is

$$\mathcal{E}_{rad}/\mathcal{E}_0 = \frac{6}{5}\frac{1}{(\Delta r)^2}, \tag{9.17}$$

independent of r. For finite r, the leading sheath carries slightly more energy than the trailing sheath, but this imbalance disappears as $r \to \infty$.

When $\Delta r \leq 2$, we must rearrange the limits in Eq. (9.16) with $r < r + \Delta r < r + 2 < r + \Delta r + 2$, and use the F-values in Eq. (5.8). In this case, we obtain contributions from all three regions. The function, F_{mo}, gives $(3/16)(2 - \Delta r)^3$. Thus, the middle region provides a maximum value of 3/2 when $\Delta r = 0$ and nothing at all when $\Delta r = 2$. The functions F_{1o} and F_{2o} give $(3/4)\Delta r\left[1 - (3/4)\Delta r + (3/20)(\Delta r)^2\right] \pm (\Delta r)^2(\Delta r - 3)^2/32r$, respectively. As $r \to \infty$, the imbalance vanishes, leaving equal contributions that range from 3/20 at $\Delta r = 2$ to 0 when $\Delta r = 0$. Thus, as Δr increases from 0 to 2, the energy distribution shifts from the middle region to the ends (which, as we have already seen, provide all the energy when $\Delta r \geq 2$). Adding all three contributions, we obtain a combined value for $\Delta r \leq 2$:

$$\mathcal{E}_{rad}/\mathcal{E}_0 = \frac{3}{2} - \frac{3}{4}\Delta r + \frac{3}{80}(\Delta r)^3, \tag{9.18}$$

independent of r.

Equations (9.17) and (9.18) and their derivatives join continuously at $\Delta r = 2$, giving the smooth, monotonically decreasing function shown in the right panel of Fig. 9.4. Thus, a sudden turn-on not only radiates the greatest amount of magnetic flux, $(3/4)\Phi_0$, but it also radiates the greatest amount of energy, $(3/2)\mathcal{E}_0$.

We can gain further insight by considering what happens when the current is turned on suddenly and then off either suddenly or gradually. As described in Chap. 4, the sudden turn-on provides 3 units of energy: 1 for the uniform field inside the sphere, 1/2 for the dipole field outside the sphere, and 3/2 radiated away. If the current were then turned off suddenly, there would be no current to interact with the transient electric field as it propagates inward and then outward through the sphere, and none of the field energy could be recovered by the source. The entire 3/2 of a unit of field energy would be radiated away in the second bubble.

However, if the current were turned off gradually, then after each increment of time, there would still be some current left to interact with the electric field of the transient (at least until the final transient is generated at the end of the ramp). This residual current allows the source to recover the energy, $(3/2) - (6/5)/(\Delta r)^2$, which would be nearly 3/2 if the turn-off were gradual enough ($\Delta r \gg 1$). Thus, in Chap. 4, it was no accident that the sudden turn-on produced equal amounts of radiated and stored energy. This is another way of saying for any field configuration that when the current is turned off, the maximum energy that can be radiated away is the amount that resides in the field and that this maximum can be attained only when the current is turned off suddenly.

Chapter 10
Radiated Energy for General Current Ramps

As discussed in Chap. 4, the Poynting flux and field energy are balanced by a source term, $-\mathbf{J}\cdot\mathbf{E}$, as given in Eq. (4.5). Thus, we can obtain the radiated energy by subtracting the field energy, $(3/2)\mathcal{E}_0$, from the total energy obtained by integrating the source term over the duration of the transient. This is computationally easier than performing a surface integral of the Poynting flux, which involves the product of the two factors, $\partial F/\partial r$ and $\partial F/\partial r_L$. On the other hand, the surface integral leads to a physically interesting way of interpreting the radiation in terms of the bends in the profile of the current ramp. Consequently, we shall describe these two approaches separately in our consideration of generalized current ramps in the next two subsections.

10.1 Source Term Approach

As mentioned above, the radiated energies in Eqs. (9.17) and (9.18) can also be obtained by integrating the source term, $-\mathbf{J}\cdot\mathbf{E}$, over the duration of the transient and the volume of the current. For this purpose, we make the following substitutions in Eq. (4.6): $f(1, r_L) \rightarrow F(1, r_L, \Delta r)$ and $J_\phi \rightarrow J_\phi\, G(r_L - 1)$, where G is the profile of the current ramp, which starts when the current begins to turn on at $r_L = 1$. In this case, Eq. (4.6) gives the total energy, $\mathcal{E}_T$, expressed as a fraction of the volume energy, $\mathcal{E}_0$, in the sphere:

$$\mathcal{E}_T/\mathcal{E}_0 = 3\int_1^{\Delta r+3} \frac{\partial F(1, r_L, \Delta r)}{\partial r_L} G(r_L - 1)\, dr_L. \tag{10.1}$$

Here, the limits, 1 and $\Delta r + 3$, correspond to the times that the leading and trailing edges cross the surface of the sphere at $r = 1$.

N. R. Sheeley, Jr., *Transient Magnetic Fields*,
https://doi.org/10.1007/978-3-030-40264-8_10

Next, we integrate by parts to obtain

$$\mathcal{E}_T/\mathcal{E}_0$$
$$= 3\left[F(1,r_L,\Delta r)G(r_L-1)|_1^{\Delta r+3} - \int_1^{\Delta r+3} F(1,r_L,\Delta r)G'(r_L-1)\,dr_L\right]. \tag{10.2}$$

Now, we recognize that $F(1, \Delta r + 3, \Delta r)G(\Delta r + 2) = 1\times1 = 1$ and that $F(1, 1, \Delta r)G(0) = 0\times0 = 0$, so that the constant term inside the brackets is just 1, and the equation becomes

$$\mathcal{E}_T/\mathcal{E}_0 = 3\left[1 - \int_1^{\Delta r+3} F(1,r_L,\Delta r)\,G'(r_L-1)\,dr_L\right]. \tag{10.3}$$

Next, we break the integral into pieces that correspond to the three domains of F (i.e., $1o$, mo, and $2o$). If $\Delta r \geq 2$, then

$$\int_1^{\Delta r+3} F\,G'\,dr_L = \int_1^3 F_{1o}\,G'\,dr_L + \int_3^{\Delta r+1} F_{mo}\,G'\,dr_L + \int_{\Delta r+1}^{\Delta r+3} F_{2o}\,G'\,dr_L. \tag{10.4}$$

Here, we can see that F_{2o} does not contribute because $G'(r_L - 1) = 0$ for $r_L > \Delta r + 1$. So the third integral vanishes when $\Delta r \geq 2$. If $\Delta r \leq 2$, then

$$\int_1^{\Delta r+3} F\,G'\,dr_L = \int_1^{\Delta r+1} F_{1o}\,G'\,dr_L + \int_{\Delta r+1}^{3} F_{mo}\,G'\,dr_L + \int_3^{\Delta r+3} F_{2o}\,G'\,dr_L. \tag{10.5}$$

For this equation, the condition that $G'(r_L - 1) = 0$ for $r_L > \Delta r + 1$ means that F_{mo} and F_{2o} do not contribute when $\Delta r \leq 2$.

With this understanding, we can introduce the variable $\xi = r_L - 1$ and simplify these expressions for the total energy. In general, the expression is

$$\mathcal{E}_T/\mathcal{E}_0 = 3\left[1 - \int_0^{\Delta r} F(1,\xi+1,\Delta r)\,G'(\xi)\,d\xi\right]. \tag{10.6}$$

The contributing f-values depend on whether Δr is larger or smaller than 2. When $\Delta r \geq 2$, we have

$$\mathcal{E}_T/\mathcal{E}_0 = 3\left[1 - \int_0^2 F_{1o}(1,\xi+1,\Delta r)\,G'(\xi)\,d\xi - \int_2^{\Delta r} F_{mo}(1,\xi+1,\Delta r)\,G'(\xi)\,d\xi\right], \tag{10.7}$$

and when $\Delta r \leq 2$, the energy is given by

$$\mathcal{E}_T/\mathcal{E}_0 = 3\left[1 - \int_0^{\Delta r} F_{1o}(1, \xi+1, \Delta r)\, G'(\xi)\, d\xi\right]. \tag{10.8}$$

In these equations, the exterior f-values, F_{1o} and F_{mo}, are obtained from Eq. (5.7) (for $\Delta r \geq 2$) and Eq. (9.2) (for $\Delta r \leq 2$) using the replacement $dr'_L/\Delta r \rightarrow G'(r_L - r'_L)dr'_L$. Substituting $\eta = r_L - r'_L$, we obtain for $\Delta r \geq 2$

$$F_{1o}(1, \xi+1, \Delta r) = \int_0^{\xi} f_o(1, \xi+1-\eta)\, G'(\eta)\, d\eta, \tag{10.9a}$$

$$F_{mo}(1, \xi+1, \Delta r) = \int_{\xi-2}^{\xi} f_o(1, \xi+1-\eta)\, G'(\eta)\, d\eta + \int_0^{\xi-2} G'(\eta)\, d\eta, \tag{10.9b}$$

where f_o is the vector potential given by Eq. (2.11a). When $\Delta r \leq 2$, we need only $F_{1o}(1, \xi+1, \Delta r)$, which is identical to the value of $F_{1o}(1, \xi+1, \Delta r)$ obtained with $\Delta r \geq 2$ in Eq. (10.9a).

With these expressions, we can now calculate the radiated energy as a function of the ramp width Δr for a variety of ramps. As an illustration, we consider the ramp profile, $G(\eta) = (\eta/\Delta r)^2$, corresponding to the QL on- ramp of Sect. 7.2. In this case, $G'(\eta) = (2/\Delta r)(\eta/\Delta r)$ and $f_o(1, \xi-\eta+1) = (1/4)\left[6(\xi-\eta) - (\xi-\eta)^3\right]$. Substituting these values in Eqs. (10.7)–(10.9) and performing the necessary integration, we find that $\mathcal{E}_T/\mathcal{E}_0 = 3 - (3/5)\Delta r + (3/140)\Delta r^3$ when $\Delta r \leq 2$ and that $\mathcal{E}_T/\mathcal{E}_0 = (3/2) + (12/5)(\Delta r)^{-2} - (72/35)(\Delta r)^{-4}$ when $\Delta r \geq 2$. In each case, the radiated energy is 3/2 of a unit less. Thus, the radiated energy starts at a value of 3/2 when $\Delta r = 0$ and decreases asymptotically as $(12/5)(\Delta r)^{-2}$ as $\Delta r \rightarrow \infty$. For comparison, the linear ramp produced a radiated energy that started at 3/2 and then decreased asymptotically as $(6/5)(\Delta r)^{-2}$ as $\Delta r \rightarrow \infty$. Later in this section, we shall see that the greater asymptotic value for the QL on-ramp is not due to the curvature (which actually reduces the contribution from the Q-end of the ramp), but instead is due to the greater slope at the L-end of the ramp.

Next, we calculate the radiated energy for the exponential current ramp shown in the left panel of Fig. 7.6. Because this exponential on-ramp is defined on the infinite domain (0,∞), only two boundaries pass the current source at $r = 1$—one when $r_L = 1$, and one when $r_L = 3$. Consequently, Eq. (10.7) applies for the full range of α, provided that the upper limit of the F_{mo} integral is changed from Δr to ∞. We obtain F_{1o} and F_{mo} by substituting $G'(\xi) = \alpha e^{-\alpha\xi}$ into Eqs. (10.9a) and (10.9b) and then use these f-values to evaluate the integrals in Eq. (10.7) (with the upper limit of the F_{mo} integral changed from Δr to ∞). Then, subtracting 3/2 of a unit of field energy from the result, we obtain the radiated energy

$$\mathcal{E}_{rad}/\mathcal{E}_0 = \left(\frac{3}{2}\right)\left[1 - (1+\alpha)\, h(\alpha)\, e^{-\alpha}\right], \tag{10.10}$$

in units of the energy, $\mathcal{E}_0$, left inside the sphere. Here, $h(\alpha) = (3/2\alpha^3)\left[(1+\alpha)\,e^{-\alpha} - (1-\alpha)\,e^{\alpha}\right]$, as defined previously in Eq. (7.16).

Because this equation applies over the full range of α, we can use it to obtain approximate expressions when $\alpha \ll 1$ and $\alpha \gg 1$. For $\alpha \ll 1$, we obtain

$$\mathcal{E}_{rad}/\mathcal{E}_0 \approx \left(\frac{3}{5}\right)\alpha^2 - \left(\frac{1}{2}\right)\alpha^3, \tag{10.11}$$

and for $\alpha \gg 1$, we find

$$\mathcal{E}_{rad}/\mathcal{E}_0 \approx \frac{3}{2} - \left(\frac{9}{4}\right)\frac{1}{\alpha} + \left(\frac{9}{4}\right)\frac{1}{\alpha^3}. \tag{10.12}$$

Thus, $1/\alpha$ is like Δr for the linear and quadratic ramps. When the decay time scale, $1/\alpha$, is small, the radiated energy is close to 3/2, but when the decay time scale is large, the energy decreases asymptotically as $(3/5)\alpha^2$.

Table 10.1 lists exact values of the radiated energies computed for the linear (LL), quadratic (QL and LQ), and symmetric cubic (CC) current profiles. Note that these exact expressions are similar to the approximate values that we obtained for the exponential on-ramp in Eqs. (10.11) and (10.12). All of these energies begin at 3/2 of a unit and decrease linearly at first. Later, the radiated energy for the linear, quadratic, and exponential profiles all decrease quadratically with coefficients of 6/5, 12/5, and 3/5, respectively. The radiated energy for the symmetric cubic has a more rapid asymptotic decay, falling as an inverse third power of the ramp length. These results are easily understood from a power series expansion of Eq. (10.7), which we compute next.

Here, the idea is to extract Δr explicitly from the ramp profile, and then expand the resulting expression for $\mathcal{E}/\mathcal{E}_0$ as a power series in the small quantity $1/\Delta r$. We do this by introducing new variables $x = \eta/\Delta r$ in Eqs. (10.9a) and (10.9b) and $y = \xi/\Delta r$ in Eq. (10.7), and then replacing $G(\xi)$ and $G(\eta)$ by new functions $H(x) = H(\xi/\Delta r) = G(\xi)$ and $H(y) = H(\eta/\Delta r) = G(\eta)$. Thus, while η and ξ range from 0 to Δr, x and y range from 0 to 1. With these substitutions, the F_{1o} and F_{mo} integrals in Eq. (10.7) become

Table 10.1 Radiated energies, $\mathcal{E}_{rad}/\mathcal{E}_0$

$G(\xi)$	$\Delta r \le 2$	$\Delta r \ge 2$
$\xi/\Delta r$	$(3/2) - (3/4)\Delta r + (3/80)(\Delta r)^3$	$(6/5)(\Delta r)^{-2}$
$(\xi/\Delta r)^2$	$(3/2) - (3/5)\Delta r + (3/140)(\Delta r)^3$	$(12/5)(\Delta r)^{-2} - (72/35)(\Delta r)^{-4}$
$1 - (\xi/\Delta r - 1)^2$	$(3/2) - (3/5)\Delta r + (3/140)(\Delta r)^3$	$(12/5)(\Delta r)^{-2} - (72/35)(\Delta r)^{-4}$
$3(\xi/\Delta r)^2 - 2(\xi/\Delta r)^3$	$(3/2) - (81/140)\Delta r + (1/56)(\Delta r)^3$	$12(\Delta r)^{-3} - (648/35)(\Delta r)^{-4}$ $+(64/7)(\Delta r)^{-6}$

$$J_{1o} = \int_0^{2/\Delta r} H'(y)\, dy \int_0^y f_o(1, 1 + (y - x)\Delta r)\, H'(x)\, dx \tag{10.13a}$$

$$J_{mo} = \int_{2/\Delta r}^{1} H'(y)\, dy \int_{y-2/\Delta r}^{y} f_o(1, 1 + (y - x)\Delta r)\, H'(x)\, dx$$
$$+ \int_{2/\Delta r}^{1} H'(y)\, dy \int_0^{y-2/\Delta r} H'(x)\, dx, \tag{10.13b}$$

where Δr now appears explicitly, and the ramp profile, G', has been replaced by the dimensionless function, H' (i.e., $H'(x) = 1$ for the linear ramp or $H'(x) = 2x$ for the QL ramp, or $H'(x) = 6x - 6x^2$ for the symmetric cubic ramp). Next, we substitute $\epsilon = 2/\Delta r$ and replace $f_o(1, 1+(y-x)\Delta r)$ by $(3/\epsilon)(y-x)-(2/\epsilon^3)(y-x)^3$ to obtain:

$$J_{1o} = \int_0^{\epsilon} H'(y)\, dy \int_0^y \left[\left(\frac{3}{\epsilon}\right)(y - x) - \left(\frac{2}{\epsilon^3}\right)(y - x)^3\right] H'(x)\, dx \tag{10.14a}$$

$$J_{mo} = \int_{\epsilon}^{1} H'(y)\, dy \int_{y-\epsilon}^{y} \left[\left(\frac{3}{\epsilon}\right)(y - x) - \left(\frac{2}{\epsilon^3}\right)(y - x)^3\right] H'(x)\, dx$$
$$+ \int_{\epsilon}^{1} H'(y)\, dy \int_0^{y-\epsilon} H'(x)\, dx. \tag{10.14b}$$

At this point, our objective is to expand these integrals as power series in the small quantity ϵ, keeping the final product accurate to $\mathcal{O}(\epsilon^3)$. For those terms that contain the factor $2/\epsilon^3$, it is necessary to retain terms of $\mathcal{O}(\epsilon^6)$ to achieve the desired accuracy. To keep the algebra manageable, we use Mathematica© software to do the x-integration, and then perform the y-integration afterwards. Substituting the result in Eq. (10.7), replacing ϵ by $2/\Delta r$, and subtracting the field energy, 3/2, we obtain:

$$\mathcal{E}_{rad}/\mathcal{E}_0 = \frac{3}{5}\frac{1}{(\Delta r)^2}\left[\{H'(0)\}^2 + \{H'(1)\}^2\right] - \frac{1}{(\Delta r)^3}\int_0^1 H'(y)H^{(3)}(y)\, dy \tag{10.15}$$

The last integral can be integrated by parts to give:

$$\mathcal{E}_{rad}/\mathcal{E}_0 = \frac{3}{5}\frac{1}{(\Delta r)^2}\left[\{H'(0)\}^2 + \{H'(1)\}^2\right]$$
$$- \frac{1}{(\Delta r)^3}\left[H'(1)H''(1) - H'(0)H''(0)\right] + \frac{1}{(\Delta r)^3}\int_0^1 [H''(y)]^2\, dy, \tag{10.16}$$

or if we return to the original ramp functions:

$$\mathcal{E}_{rad}/\mathcal{E}_0 = \frac{3}{5}\left[\{G'(0)\}^2 + \{G'(\Delta r)\}^2\right] - \left[G'(\Delta r)G''(\Delta r) - G'(0)G''(0)\right] + \int_0^{\Delta r} [G''(\xi)]^2\, d\xi \tag{10.17}$$

The problem is much easier if $\Delta r \leq 2$. In this case, we use Eq. (10.8) with F_{1o} given by Eq. (10.9a). The corresponding value of J_{1o} is

$$J_{1o} = \int_0^1 H'(y)\,dy \int_0^y f_o(1, 1 + (y - x)\Delta r) H'(x)\,dx, \tag{10.18}$$

where $f_o(1, 1 + (y - x)\Delta r) = (3\Delta r/2)(y - x) - \{(\Delta r)^3/4\}(y - x)^3$. It follows immediately that there are only two terms in the expression for J_{1o}:

$$J_{1o} = \left(\frac{3I_1}{2}\right)(\Delta r) - \left(\frac{I_2}{4}\right)(\Delta r)^3, \tag{10.19}$$

where I_1 and I_2 are given by

$$I_1 = \int_0^1 H'(y)dy \int_0^y (y - x)\, H'(x)dx = \int_0^1 \left\{H(y) - H^2(y)\right\} dy, \tag{10.20a}$$

$$I_2 = \int_0^1 H'(y)dy \int_0^y (y - x)^3\, H'(x)dx = 3\int_0^1 H'(y)dy \int_0^y (y - x)^2\, H(x)dx. \tag{10.20b}$$

Consequently, the radiated energy is

$$\mathcal{E}_{rad}/\mathcal{E}_0 = \frac{3}{2} - \left(\frac{9I_1}{2}\right)(\Delta r) + \left(\frac{3I_2}{4}\right)(\Delta r)^3, \tag{10.21}$$

where I_1 and I_2 are positive quantities for a monotonically increasing current ramp. Together, Eqs. (10.16) and (10.21) give the radiated energy for a general current ramp that rises from 0 to 1 in a time Δr. The difference is that Eq. (10.21) is an exact expression, whereas Eq. (10.16) is a series approximation, valid through terms of $\mathcal{O}(1/(\Delta r)^3)$.

In general, the asymptotic expression given by Eq. (10.16) will not match the exact expression given by Eq. (10.21) when $\Delta r = 2$. Consequently, to obtain a continuous distribution of radiated energy, it will be necessary to use Eqs. (10.14a) and (10.14b) to calculate J_{1o} and J_{mo} exactly for $\Delta r \geq 2$, and then deduce the radiated energy from Eq. (10.7) in the form $\mathcal{E}_{rad}/\mathcal{E}_0 = 3\,(1/2 - J_{1o} - J_{mo})$.

In particular, a mismatch is obtained for the sinusoidal profile, $H(x) = \sin^2(\pi x/2)$. For $\Delta r \leq 2$, the radiated energy is easily obtained from Eqs. (10.20) and (10.21). The value is $\mathcal{E}_{rad}/\mathcal{E}_0 = 3/2 - (9/16)(\Delta r) + (3/16)\left\{1 - (3/\pi)^2\right\}(\Delta r)^3$.

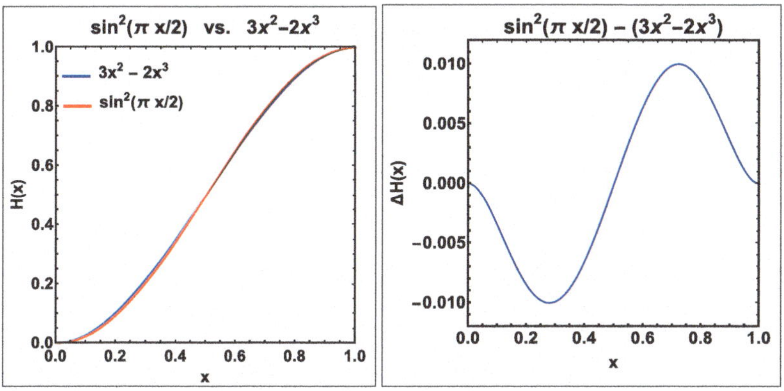

Fig. 10.1 A verification that the sinusoidal current ramp, $H(x) = \sin^2(\pi x/2)$, is closely approximated by the symmetric cubic profile, $H(x) = 3x^2 - 2x^3$. (Left) The two profiles plotted together; (right) a plot of their difference

However, the asymptotic value is $\mathcal{E}_{rad}/\mathcal{E}_0 = (\pi^4/8)\,(\Delta r)^{-3}$, which is 1.522 at $\Delta r = 2$ and is not close to the value of 0.507 obtained from the exact expression when $\Delta r = 2$. To gain good agreement, the sinusoidal expression must be expanded to include more terms. In particular, terms of $\mathcal{O}\{(\Delta r)^{-12}\}$ are necessary to obtain agreement to three decimal places. On the other hand, as shown in Fig. 10.1, this sinusoidal profile is well fit by the symmetric cubic profile, $H(x) = 3x^2 - 2x^3$, so that the symmetric cubic solution given in Table 10.1 may be applied to the sinusoidal profile. Likewise, the space-time maps and plots of flux contours for the symmetric cubic will also give accurate representations for the sinusoidal profile. This near equivalence will be of some interest when we consider oscillating currents in Chap. 11.

From Eq. (10.16) we can now understand the asymptotic dependences of the radiated energy in terms of the initial and final slopes of the ramp profiles, which are the source of the $1/(\Delta r)^2$-dependence. The linear ramp has a slope $H'(x) = 1$. In particular, $H'(0) = 1$ and $H'(1) = 1$ and both ends contribute to the radiated energy, giving $2\times(3/5)/(\Delta r)^2 = (6/5)/(\Delta r)^2$. The quadratic ramp QL has only one linear end, but the slope there is $H'(1) = 2$, giving 4 times the amount for a unit slope: $4\times(3/5)/(\Delta r)^2 = (12/5)/(\Delta r)^2$. The same result occurs for the other quadratic ramp LQ. If we associate the exponential decay rate, α, with a decay time $\Delta r = 1/\alpha$, then we can do a similar calculation for that profile. In this case, the initial slope is $H'(0) = 1$, and there is no contribution from the distant end of the ramp. Consequently, the asymptotic dependence is $1\times(3/5)/(\Delta r)^2 = (3/5)/(\Delta r)^2$, or $(3/5)\alpha^2$ in terms of the decay rate α. Finally, the slope of the symmetric cubic on-ramp vanishes at both ends, so that this profile does not give a $1/(\Delta r)^2$ asymptotic dependence.

Figure 10.2 shows the radiated energy for these current ramps as given in Table 10.1 and by Eq. (10.10) for the exponential on-ramp (with $\Delta r = 1/\alpha$).

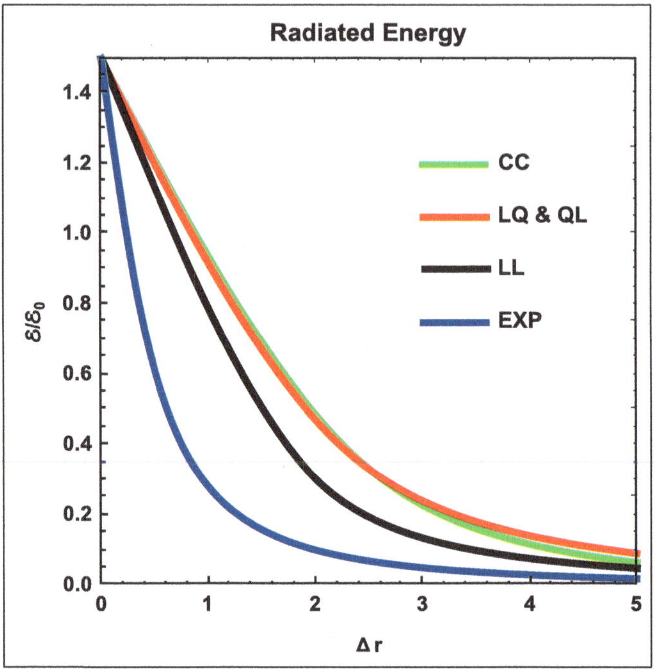

Fig. 10.2 Radiated energy for the linear on-ramp (LL), the quadratic on-ramps (LQ and QL), the exponential on-ramp (EXP) and cubic on-ramp (CC) on-ramp, plotted as a function of the ramp width (or reciprocal decay rate for the exponential). The energy is expressed in units of $\mathcal{E}_0 = (\frac{4}{3}\pi R^3)(B_0^2/2\mu_0)$, the energy inside the uniformly magnetized sphere of radius R and field strength B_0. The radiated energy begins at 3/2 a unit and decreases almost linearly at first, with the exponential showing the fastest initial decrease and the symmetric cubic showing the slowest initial decrease

For $\Delta r > 3$, the asymptotic decreases of the quadratic (red), linear (black), and exponential (blue) ramps are visible with the steeper decrease for the symmetric cubic (green) shown cutting across them toward the outer edge of the plot. However, for $\Delta r < 3$, the symmetric cubic shows the smallest rate of decrease, and its radiated energy curve nearly coincides with that of the quadratic profiles. This is consistent with the coefficients of the linear terms in Table 10.1 (and in Eq. (10.12) for the exponential on-ramp); these coefficients are approximately the same for the symmetric cubic (0.58) and the quadratic profiles (0.60) and are larger for the linear ramp (0.75) and the exponential (2.25).

This suggests that the initial decrease would be even less for a ramp whose central slope is higher. For example, the symmetric fifth-power profile, $H(x) = 10x^3 - 15x^4 + 6x^5$, has a central slope of $H'(1/2) = 15/8 = 1.875$, which is greater than the value of 1.5 obtained for the symmetric cubic. Calculations for this fifth-power profile give $\mathcal{E}_{rad}/\mathcal{E}_0 = 3/2 - (75/154)\Delta r + \mathcal{O}(\Delta r)^3$ for $\Delta r \leq 2$, corresponding to an initial decay rate of $75/154 \approx 0.49$, and $\mathcal{E}_{rad}/\mathcal{E}_0 \approx (120/7)/(\Delta r)^3$ for $\Delta r \geq 2$.

Thus, the initial decay rate is smaller than that of the symmetric cubic, and the asymptotic value of $(120/7)/(\Delta r)^3 \approx 17.1/(\Delta r)^3$ is higher than the corresponding value for the symmetric cubic, consistent with the trend in Table 10.1 and Fig. 10.2.

Another approach is to rewrite the expression for I_1 as

$$I_1 = 1/4 - \int_0^1 [H(y) - 1/2]^2 \, dy = 1/4 - \int_0^1 K^2(y) \, dy, \tag{10.22}$$

where $K(y) = H(y) - 1/2$. Whereas $H(y)$ ranges from 0 to 1, $K(y)$ ranges from $-1/2$ to $+1/2$, and $K^2(y)$ ranges from 0 to $+1/4$. Thus, to obtain the least initial decay, we need to find a profile that makes I_1 small and therefore keeps $K^2(y)$ close to $+1/4$ over most of the range of y. We can do that using a simple analytical expression for $H(y)$:

$$H(y) = \begin{cases} \frac{1}{2}(2y)^n, & 0 \le y \le 1/2 \\ 1 - \frac{1}{2}\left[2(1-y)\right]^n, & 1/2 \le y \le 1, \end{cases} \tag{10.23}$$

where n is an index indicating the steepness of the profile. In this case,

$$H'(y) = K'(y) = \begin{cases} n(2y)^{n-1}, & 0 \le y \le 1/2 \\ n\left[2(1-y)\right]^{n-1}, & 1/2 \le y \le 1, \end{cases} \tag{10.24}$$

and we can see that $n = H'(1/2) = K'(1/2)$ is the central slope of the two profiles. Using Eq. (10.22) to evaluate I_1, we obtain

$$I_1 = \frac{(3n+1)}{4(n+1)(2n+1)}. \tag{10.25}$$

Consequently, for $n \gg 1$, $I_1 \approx 3/(8n)$ so that $\mathcal{E}_{rad} \approx 3/2 - (27/16n)(\Delta r) + \mathcal{O}[(\Delta r)^3]$, and the initial rate of decay is $d\mathcal{E}_{rad}/d\Delta r \approx -27/16n$ when $\mathcal{E}_{rad}$ is in units of $\mathcal{E}_0$.

Figure 10.3 shows plots of $H(y)$, $K(y)$, and $K^2(y)$ when the index $n = 9$. In this case, the maximum slope $K'(1/2) = 9$, and $\int_0^1 K^2(y)dy$ is given by the shaded area in the right panel. Because the total area is 1/4, we can see from Eq. (10.16) that the sliver of unshaded area is equal to the value of I_1, which Eq. (10.21) shows is a measure of the initial decrease of radiated energy with respect to Δr. Therefore, a small initial decrease corresponds to a small sliver of unshaded area and a steep slope of the current ramp. As $n \to \infty$, the profile approaches a step function and the radiated energy approaches $(3/2)\mathcal{E}_0$.

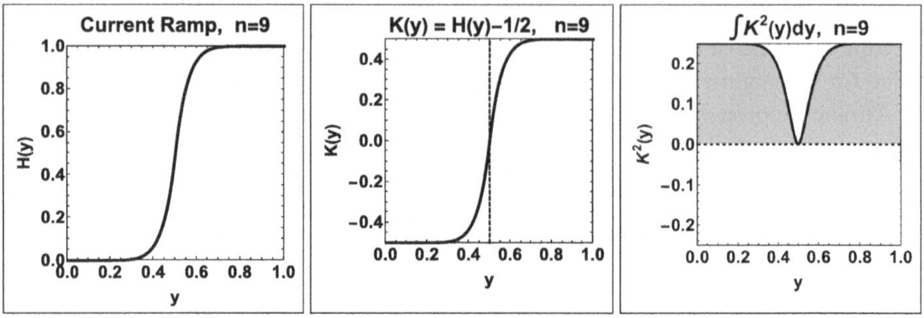

Fig. 10.3 Current ramp profile, $H(y)$ (left), profile of $K(y) = H(y) - 1/2$ (center), and $K^2(y)$ (right), plotted for a slope index $n = 9$. In the right panel, the shaded area represents $\int_0^1 K^2(y)\,dy$, and the sliver of unshaded area corresponds to I_1 in Eqs. (10.20a) and (10.21), and is a measure of the initial rate of decrease of radiated energy with respect to the ramp width, Δr. Thus, a steep profile provides the greatest amount of radiated energy up to the limit of $(3/2)\mathcal{E}_0$ for a sudden onset

10.2 Poynting Vector Approach

In this section, we evaluate the radiated energy directly from Eq. (9.16), which expresses the energy in terms of contributions from the beginning ($1o$), middle (mo), and end ($2o$) of the transient. As we shall see, this approach provides an alternative way of understanding the radiation as a running average of the current profile.

Each term of Eq. (9.16) consists of an integral of a two-factor product of the form $(\partial F/\partial r)(\partial F/\partial r_L)$. Thus, for the middle region, we have

$$\frac{\mathcal{E}_{rad}}{\mathcal{E}_0}\Big|_{mo} = -\frac{1}{r^2}\int_{r+2}^{r+\Delta r} \frac{\partial F_{mo}}{\partial r}\frac{\partial F_{mo}}{\partial r_L}\,dr_L. \tag{10.26}$$

In Chap. 7, we derived $\partial F_{mo}/\partial r$ from the value of F_{mo} given in Eq. (5.7). The result is given in Eq. (7.3), which we rewrite here as

$$\frac{\partial F_{mo}(r, r_L)}{\partial r} = \int_{r_L-(r+2)}^{r_L-r} \frac{\partial f_o(r, r_L-\xi)}{\partial r}\,G'(\xi)\,d\xi. \tag{10.27}$$

Next, taking f_o from Eq. (2.11a), we can obtain $\partial f_o(r, r_L - \xi)/\partial r = (3r/2)(r_L - \xi - 1 - r)$, and substitute this value in Eq. (10.27) to obtain

$$\frac{\partial F_{mo}(r, r_L)}{\partial r} = \frac{3r}{2}\int_{r_L-r-2}^{r_L-r} (r_L - \xi - 1 - r)\,G'(\xi)\,d\xi = \left(\frac{3r}{2}\right) I_{mo}, \tag{10.28}$$

where I_{mo} is given by

$$I_{mo} = \int_{r_L-r-2}^{r_L-r} (r_L - \xi - 1 - r)\, G'(\xi)\, d\xi$$
$$= \int_{r_L-r-2}^{r_L-r} G(\xi)\, d\xi - [G(r_L - r) + G(r_L - r - 2)]\,. \tag{10.29}$$

Next, defining $v = (r_L - r)/\Delta r$ and introducing $x = \xi/\Delta r$, we obtain

$$\frac{I_{mo}}{\Delta r} = \int_{v-\frac{2}{\Delta r}}^{v} H(x)\, dx - \left(\frac{2}{\Delta r}\right)\left[\frac{H(v) + H\left(v - \frac{2}{\Delta r}\right)}{2}\right], \tag{10.30}$$

where $H(x) = H(\xi/\Delta r) = G(\xi)$.

A similar procedure with $\partial F_{mo}/\partial r_L$ gives

$$\frac{\partial F_{mo}(r, r_L)}{\partial r_L} = \int_{r_L-(r+2)}^{r_L-r} \frac{\partial f_o(r, r_L - \xi)}{\partial r_L}\, G'(\xi)\, d\xi$$
$$= \frac{3}{4}\int_{r_L-r-2}^{r_L-r} \left[1 + r^2 - (r_L - \xi - 1)^2\right] G'(\xi)\, d\xi. \tag{10.31}$$

Again, we define $v = (r_L - r)/\Delta r$ and introduce $x = \xi/\Delta r$. This time we obtain

$$\frac{\partial F_{mo}(r, r_L)}{\partial r_L} = \frac{3}{4}\int_{v-\frac{2}{\Delta r}}^{v} [1 + (x\Delta r - v\Delta r + 1)(2r - x\Delta r + v\Delta r - 1)]\; H'(x)\, dx. \tag{10.32}$$

Here, we retain only the terms of order r (whose product offsets the $-1/r^2$ factor in Eq. (10.26)), and obtain

$$\frac{\partial F_{mo}(r, r_L)}{\partial r_L} = \frac{3r}{2}\int_{v-\frac{2}{\Delta r}}^{v} [1 + (x - v)\Delta r]\; H'(x)\, dx = -\left(\frac{3r}{2}\right) I_{mo}, \tag{10.33}$$

where I_{mo} is the same quantity that is given in Eq. (10.30). Referring back to Eq. (10.26), it follows that

$$\frac{\mathcal{E}_{rad}}{\mathcal{E}_0}\Big|_{mo} = \frac{9}{4}\Delta r \int_{\frac{2}{\Delta r}}^{1} I_{mo}^2\, dv = \frac{9}{4}(\Delta r)^3 \int_{\frac{2}{\Delta r}}^{1} (\Delta A_{mo})^2\, dv, \tag{10.34}$$

where ΔA_{mo} is the area given by

$$\Delta A_{mo} = \int_{v-\frac{2}{\Delta r}}^{v} H(x)\, dx - \left(\frac{2}{\Delta r}\right)\left[\frac{H(v) + H\left(v - \frac{2}{\Delta r}\right)}{2}\right] \tag{10.35}$$

Thus, for each point from $v = 2/\Delta r$ to $v = 1$, ΔA_{mo} is the area between a region of the current ramp between v and $v - 2/\Delta r$ and the line segment joining the

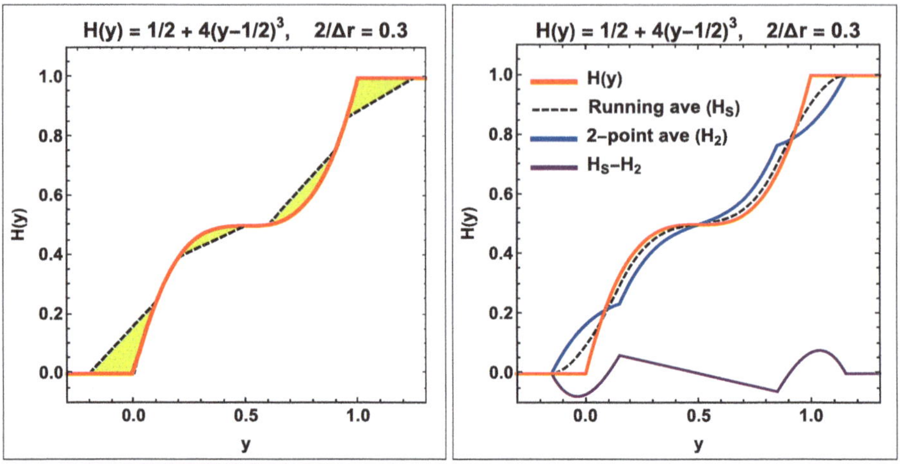

Fig. 10.4 Line plots, illustrating the way that the bends of a cubic current ramp contribute to the radiated energy. (Left) Dashed line segments draped behind four points along the profile, mark yellow areas, ΔA, that contribute to the radiated energy. (Right) An equivalent treatment using running averages to obtain profile differences, $H_S - H_2$, that are related to the areas, ΔA, via the equation $H_S - H_2 = (\Delta r/2)\Delta A$

ends of this region, as illustrated in the left panel of Fig. 10.4. In this case, we are referring to the two middle areas of the figure.

The two areas near the ends of the ramp refer to contributions from the $1o$ and $2o$ ends. In fact, similar mathematical derivations for those regions give

$$\Delta A_{1o} = \int_0^v H(x)\,dx - \left(\frac{2}{\Delta r}\right)\left[\frac{H(v)}{2}\right] \tag{10.36a}$$

$$\Delta A_{2o} = \int_{v-\frac{2}{\Delta r}}^{1} H(x)\,dx - \left(\frac{2}{\Delta r}\right)\left[\frac{H(v-\frac{2}{\Delta r})+1}{2}\right] + (v-1) \tag{10.36b}$$

and

$$\frac{\mathcal{E}_{rad}}{\mathcal{E}_0}|_{1o} = \frac{9}{4}(\Delta r)^3 \int_0^{\frac{2}{\Delta r}} (\Delta A_{1o})^2\,dv \tag{10.37a}$$

$$\frac{\mathcal{E}_{rad}}{\mathcal{E}_0}|_{2o} = \frac{9}{4}(\Delta r)^3 \int_1^{1+\frac{2}{\Delta r}} (\Delta A_{2o})^2\,dv. \tag{10.37b}$$

We can combine Eqs. (10.34)–(10.37) by extending the definition of the current ramp beyond the two ends so that $H(y) = 0$ for $y \le 0$ and $H(y) = 1$ for $y \ge 1$. In this case, we can introduce a new variable $u = v - 1/\Delta r$ to symmetrize the running averages and obtain the radiated energy from the general expression

$$\frac{\mathcal{E}_{rad}}{\mathcal{E}_0} = \frac{9}{4}(\Delta r)^3 \int_{-\frac{1}{\Delta r}}^{1+\frac{1}{\Delta r}} (\Delta A)^2 \, du \tag{10.38}$$

with ΔA defined by

$$\Delta A = \int_{u-\frac{1}{\Delta r}}^{u+\frac{1}{\Delta r}} H(x) \, dx - \left(\frac{2}{\Delta r}\right) \left[\frac{H\left(u - \frac{1}{\Delta r}\right) + H\left(u - \frac{1}{\Delta r}\right)}{2}\right]. \tag{10.39}$$

The result is illustrated in the right panel of Fig. 10.4. In red, we have plotted a ramp function $H(y) = 1/2 + 4(y - 1/2)^3$ on the interval (0,1) and assigned it the value 0 for smaller values of y and 1 for larger values. We plot H_S, the running average of $H(y)$ on the interval $(y - 1/\Delta r,\ y + 1/\Delta r)$, as the black dashed curve, and we plot H_2, the 2-point running average on the same interval, as the solid blue curve. Their difference, $H_S - H_2 = (\Delta r/2)\Delta A$, is shown as the purple curve toward the bottom of the panel. The radiated energy is obtained from the square of this difference as

$$\frac{\mathcal{E}_{rad}}{\mathcal{E}_0} = 9 \, \Delta r \int_{-\frac{1}{\Delta r}}^{1+\frac{1}{\Delta r}} (H_S - H_2)^2 \, dy. \tag{10.40}$$

It is instructive to expand the expression for ΔA in Eq. (10.39) as a power series in $1/\Delta r$. Including terms through order $1/(\Delta r)^3$, we find that

$$\Delta A = -\frac{2}{3} \, H''(u) \left(\frac{1}{\Delta r}\right)^3. \tag{10.41}$$

This means that the radiative energy given by Eq. (10.38) becomes

$$\frac{\mathcal{E}_{rad}}{\mathcal{E}_0} = \left(\frac{1}{\Delta r}\right)^3 \int_{-\frac{1}{\Delta r}}^{1+\frac{1}{\Delta r}} [H''(u)]^2 \, du = \left(\frac{1}{\Delta r}\right)^3 \int_0^1 [H''(y)]^2 \, dy, \tag{10.42}$$

consistent with the central contribution to the radiated energy that we obtained in the previous section, as given in Eq. (10.16). Thus, when $\Delta r \gg 2$, the contribution from the middle of the transient varies as $(\Delta r)^{-3}$ and is proportional to the square of the "acceleration," like the radiation from an accelerating charge in the non-relativistic limit (Leighton 1959).

Equation (10.41) also implies that $H_S - H_2 = -H''(y)/\{3(\Delta r)^2\}$. Substituting the values of $H''(y) = 24(y - 1/2)$ and $\Delta r = 2/0.3$, we obtain $H_S - H_2 = -0.18(y - 0.5)$, as the equation for the middle portion of the purple curve in the right panel of Fig. 10.4. So it is no coincidence that this section of the curve looks like a straight line. Of course, it would not be a straight line if the ramp profile, $H''(y)$, were a non-linear function of y.

It is interesting to see how Eqs. (10.38) and (10.39) apply to some other current ramps. A simple example is the sudden turn-on that jumps from 0 to 1 at the midpoint of the interval. As the observation window encounters the jump, there are opposite contributions from regions above and below the jump. These contributions cancel when the window is centered on the jump, but for an arbitrary offset distance x, they give $\Delta A = x$. Therefore, $(\Delta A)^2 = x^2$ for x in the interval $(-1/\Delta r, +1/\Delta r)$ and $(\Delta A)^2 = 0$ otherwise. In this case, the integral in Eq. (10.38) has the value $(2/3)(\Delta r)^{-3}$, and $\mathcal{E}_{rad}/\mathcal{E}_0 = 3/2$, as expected for a sudden onset.

The linear ramp provides another example. The running window sees no excess area from the middle part of the ramp, so we need only consider the ends. When the start of the ramp lies at a location x in the observation window, the excess area is $\Delta A = (x/2)(2/\Delta r - x)$. Therefore, $(\Delta A)^2 = (x/2)^2\ (2/\Delta r - x)^2$ for x in the range $(0, 2/\Delta r)$ and $(\Delta A)^2 = 0$ otherwise. For this range of integration, we obtain $\mathcal{E}_{rad}/\mathcal{E}_0|_{1o} = (3/5)(\Delta r)^{-2}$ from the start of the ramp. An equal contribution is obtained from the end of the ramp, so that the total radiated energy is $(6/5)(\Delta r)^{-2}$ when $\Delta r \geq 2$, in agreement with Eq. (9.17). When $\Delta r \leq 2$, the running window becomes wider than the ramp, and all three regions contribute to the radiated energy, giving $\mathcal{E}_{rad}/\mathcal{E}_0 = 3/2 - (3/4)(\Delta r) + (3/80)(\Delta r)^3$, in agreement with Eq. (9.18).

We can gain further insight by considering profiles that do not change rapidly on the interval $(0, \Delta r)$. Near the start of a very long ramp with $\Delta r \gg 2$, we may approximate $H(x)$ by the first non-vanishing term in its power series expansion about $x = 0$:

$$H(x) = \frac{H^{(k)}(0)}{k!}\ x^k. \tag{10.43}$$

Substituting this expression into Eq. (10.39), evaluating the integral, and then integrating the resulting squared area over the range $(0, 2/\Delta r)$, we obtain

$$\frac{\mathcal{E}_{rad}}{\mathcal{E}_0}|_{1o} = (\Delta r)^{-2k} \left[\frac{H^{(k)}(0)}{k!}\right]^2 C(k), \tag{10.44}$$

where $C(k)$ is given by

$$C(k) = 9\ \frac{2^{(2k-1)}\ (2k^3 - k^2 + 1)}{(k+1)^2\ (2k+1)(2k+3)}. \tag{10.45}$$

Here, the profile index, k, is 1 for a linear ramp, 2 for a quadratic ramp, and so forth. A similar expression is obtained for the $2o$ end of the ramp, except that $H^{(k)}(0)$ is replaced by $H^{(k)}(1)$, and k will have a different value than for the $1o$ region unless the "acceleration" at the start of the ramp is the same as the "deceleration" at the end. The important point here is that the radiated energy varies asymptotically as $(\Delta r)^{-2k}$.

Together, Eqs. (10.42), (10.44), and the analogous relation for the 2o region help us to understand the asymptotic contributions to the radiated energy. For a curved ramp, the contribution from the middle of the ramp is $\mathcal{E}_{mo}/\mathcal{E}_0 \sim (\Delta r)^{-3}$, but the contributions from the ends of the ramp may vary as different powers of $1/\Delta r$, depending on the amount of curvature there. In general, $\mathcal{E}_{1o}/\mathcal{E}_0 \sim (\Delta r)^{-2k}$ and $\mathcal{E}_{1o}/\mathcal{E}_0 \sim (\Delta r)^{-2j}$. Thus, for a linear ramp where $j = k = 1$ and $H''(x) = 0$, one obtains identical contributions from the ends and no contribution from the middle, giving $\mathcal{E}_{rad}/\mathcal{E}_0 = (6/5)\,(\Delta r)^{-2}$, in agreement with Eq. (9.17).

If the ramp has quadratic curvature at one end, say at $x = 0$, then $k = 2$ and $\mathcal{E}_{1o}/\mathcal{E}_0 \sim (\Delta r)^{-4}$, compared to the $(\Delta r)^{-3}$ contribution from the middle and the $(\Delta r)^{-2}$ contribution from the other end (where the slope is assumed to be linear). Therefore, the linear end dominates when $\Delta r \gg 2$, and $\mathcal{E}_{rad}/\mathcal{E}_0 \sim (3/5)\,[H'(1)/\Delta r]^2 = (3/5)\,[G'(\Delta r)]^2$, which reproduces the leading terms of order $(\Delta r)^{-2}$ in Table 10.1 and in Eqs. (10.16) and (10.17). However, if both ends are curved, then their contributions will fall off at least as fast as $(\Delta r)^{-4}$, and the $(\Delta r)^{-3}$ dependence of the middle region will dominate. In this case, we can use Eq. (10.42) to reproduce the leading energy terms for the doubly curved cubic profile in Table 10.1.

Even when the middle region dominates, the radiation does not come from the center of that region. As Eq. (10.42) indicates, the distribution is given by $[H''(y)]^2$ when $\Delta r \gg 2$ and is doubly peaked in the simple cases. A convenient illustration is provided by the symmetric ramp profile

$$H'(x) = a\,[x(1-x)]^n, \tag{10.46}$$

where a is a normalizing factor given by

$$\frac{1}{a} = \int_0^1 [u(1-u)]^n\,du = \frac{(n!)^2}{(2n+1)!}. \tag{10.47}$$

When $n = 0$, this profile is familiar to us as the linear ramp. However, for $n > 0$, its slope vanishes at each end of the ramp and obtains a maximum value at the center where $x = 1/2$.

The left panel of Fig. 10.5 shows a sample of profiles plotted for $n = 0 - 7$, starting at $n = 0$ as the dashed line. As one can see, the central slope increases with n, eventually causing the profile to approach that of the sudden turn-on.

The right panel shows the corresponding profiles of the radiated energy, $(9/4)(\Delta r)^3(\Delta A)^2$, as given in Eq. (10.38) with $\Delta r = 10$, again using a dashed line for the profile with $n = 0$. The next curve with $n = 1$ corresponds to the symmetric cubic that we encountered in Sect. 7.4. Its contributions to the radiated energy peak at the ends of the ramp and decrease toward the center, as we can see in this figure. However, for $n > 2$, the peaks occur further inside the ramp. We can calculate the locations of the maxima by setting $d\Delta A/dx = 0$. When $\Delta r \gg 2$, we can use Eq. (10.41) for ΔA, in which case the condition becomes $H^{(3)}(x) = 0$. Starting with $H'(x)$ from Eq. (10.46), we find that

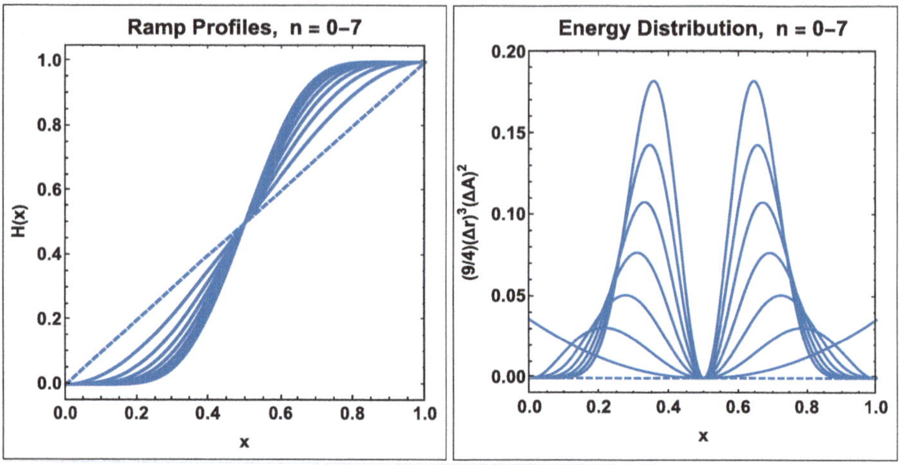

Fig. 10.5 (Left) Current ramps with slopes of the form $H'(x) \sim [x(1-x)]^n$, which vanishes at the ends of the ramp and is maximum at the center if $n > 0$; (right) the corresponding profiles of radiated energy, $(9/4)(\Delta r)^3(\Delta A)^2$, as indicated in Eqs. (10.37)–(10.38). The energy has a bimodal distribution whose peaks move centerward as n increases. In each panel, the dashed line refers to the linear ramp with $n = 0$

$$x = \frac{1}{2} \pm \frac{1}{2}(2n-1)^{-1/2}. \tag{10.48}$$

Thus, the peaks are spaced at equal distances of $(1/2)(2n-1)^{-1/2}$ on each side of the midpoint where $x = 1/2$. However, this relation eventually breaks down as n becomes large and the separation, $(2n-1)^{-1/2}$, approaches $2/\Delta r$—the width of the window used in the running average of $H(x)$, which means $n \sim (1/8)(\Delta r)^2$.

Chapter 11
Oscillating Currents

Here, we shall consider what happens when the current is repeatedly turned on and off. We begin by ramping the current up linearly during a time $\Delta r \geq 2$ (in units of R/c) and then (after an elapsed time of $2R/c$ for the bubble to leave the sphere) ramping it down again over an equal time Δr. During the turn-on, a total energy of $3/2 + (6/5)(\Delta r)^{-2}$ units is required for the linear ramp: 1 unit for the sphere, 1/2 a unit for the external dipole field, and $(6/5)\ (\Delta r)^{-2}$ of a unit for the escaping bubble. During the turn-off, $3/2 - (6/5)(\Delta r)^{-2}$ is regained from the field, but $(6/5)(\Delta r)^{-2}$ escapes with the bubble. Thus, in the process of gradually turning the current on, waiting momentarily, and then gradually turning it off again, a total energy of $(12/5)(\Delta r)^{-2}$ is lost to the escaping bubbles.

Imagine that this process were repeated indefinitely to form a sawtooth current profile whose slightly blunted teeth were separated by a period $2(\Delta r + 2)(R/c)$. Then there would be an energy loss rate or radiated power, p, given by

$$\frac{p}{p_0} = \frac{6}{5} \frac{1}{(\Delta r)^2 (\Delta r + 2)}, \tag{11.1}$$

where $p_0 = \mathcal{E}_0 c/R$. For $\Delta r \gg 2$, this loss rate would approach $p/p_0 = (6/5)(\Delta r)^{-3}$.

11.1 Overlapping Sheaths

Next, consider what would happen if we "sharpened" the teeth by beginning the downward ramp just as soon as the upward ramp reached its peak. In this case, a

Electronic Supplementary Material The online version of this chapter (https://doi.org/10.1007/978-3-030-40264-8_11) contains supplementary material, which is available to authorized users.

N. R. Sheeley, Jr., *Transient Magnetic Fields*,
https://doi.org/10.1007/978-3-030-40264-8_11

turn-off bubble would start to form before the trailing edge of the turn-on bubble had left the sphere. As the turn-off bubble began to move out of the sphere, its leading edge would overlap with the trailing edge of the turn-on bubble. Because the meridional fields in these trailing and leading "shells" have the same sign, they would interfere constructively, causing more energy to be radiated than in the non-overlapping case.

To calculate this effect, let us suppose that the current is turned on at time 0, that it increases for an interval Δr, that it is turned off at time δ, where δ lies in the interval $(\Delta r, \Delta r+2)$, and that the subsequent decrease takes an additional time Δr. Here, the upper limit $\delta = \Delta r+2$ corresponds to the case of non-overlapping bubbles considered previously, and the lower limit $\delta = \Delta r$ corresponds to the fully sawtooth profile. (If δ were smaller than Δr, the off-ramp would begin before the on-ramp ended, producing another sawtooth profile with smaller values of ramp time and height.)

For this range of δ and a repeating sequence of turn-on's and turn-off's, it is sufficient to consider the overlap between the trailing edge of one turn-on bubble and the leading edge of the next turn-on bubble. In this case, the radiated energy per "half-period," δ, may be written as

$$\frac{\mathcal{E}}{\mathcal{E}_0} = -\frac{1}{r^2}\left[\int_{r+\Delta r}^{r+\delta} \frac{\partial F_{2o}}{\partial r}\frac{\partial F_{2o}}{\partial r_L}\,dr_L + \int_{r+\delta}^{r+\Delta r+2} \frac{\partial F_3}{\partial r}\frac{\partial F_3}{\partial r_L}\,dr_L \right. \\ \left. + \int_{r+\Delta r+2}^{r+\delta+2} \frac{\partial F_{1o}}{\partial r}\frac{\partial F_{1o}}{\partial r_L}\,dr_L\right], \tag{11.2}$$

where $F_{1o} = F_{1o}(r, r_L - \delta)$ refers to the leading sheath of the off-bubble, $F_{2o} = F_{2o}(r, r_L)$ refers to the trailing sheath of the on-bubble, and $F_3(r, r_L) = F_{2o}(r, r_L) - F_{1o}(r, r_L - \delta)$ refers to the overlap. F_{mo} does not need to be included here because it gives no radiated energy for the linear ramp. Using the F-values given by Eq. (5.9) and dividing the energy by δ, we obtain the radiated power for the periodic sawtooth:

$$\frac{p}{p_0} = \frac{3}{80\delta(\Delta r)^2}\left[32 + (2 + \Delta r - \delta)^3\left\{4 + 6(\delta - \Delta r) + (\delta - \Delta r)^2\right\}\right]. \tag{11.3}$$

From Eq. (11.3), we see that when $\delta = \Delta r + 2$, the radiated power is $p/p_0 = (6/5)(\Delta r)^{-2}/(\Delta r+2)$, in agreement with Eq. (11.1). For $\Delta r \gg 2$, this expression reduces to $(6/5)(\Delta r)^{-3}$. By comparison, when $\delta = \Delta r$, Eq. (11.3) gives $p/p_0 = (12/5)(\Delta r)^{-3}$. Thus, in the long-wavelength limit, twice as much power is radiated when the current profile is fully sawtooth than when it is blunted.

Figure 11.1 compares the flux contours for a blunted profile (left) with those for a sharpened profile (right). Here, we have introduced the quantity ϵ, defined by $\delta = \Delta r + 2 - \epsilon$ to give the amount of shift from the blunted profile. Thus, when $\epsilon = 0$, the profile is blunted and when $\epsilon = 2$, the profile is sharpened. In each case, the ramp width is $\Delta r = 5$. In the left panel, the fields of the blunted profile show

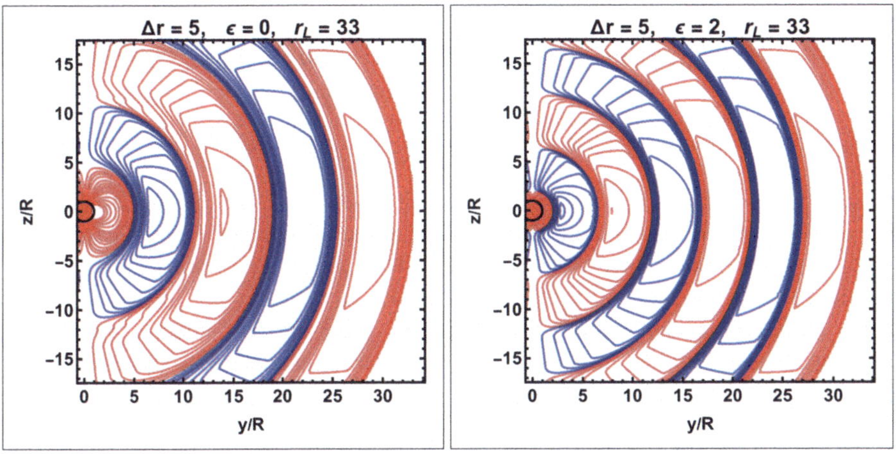

Fig. 11.1 (Left) The flux contours for a linear on-ramp of width $\Delta r = 5$, made periodic by waiting 2 units to allow the first bubble to exit the sphere before the next one begins (giving the "blunted ramp" with $\epsilon = 0$). (Right) The corresponding contours when the off-ramp begins as soon as the on-ramp reaches its maximum value of 1 (giving the "sharpened ramp" with $\epsilon = 2$). For the sharpened ramp, the waves are closer together and the sheath fields are fully overlapped, giving a greater average radiated power than for the blunted ramp. As in our previous figures, red indicates clockwise circulation of the flux contours and blue indicates counterclockwise circulation, so that adjacent red and blue fields are pointed in the same direction

alternating on-bubbles and off-bubbles beginning with the leading edge of the first on-bubble at $r_L = 33$ and continuing to the third on-bubble moving out from the sphere. The red and blue colors refer to the circulation within each bubble, which causes the adjacent fields of red and blue to point in the same direction.

In the right panel, these regions fully overlap, so that the 2-unit width now contains flux from the trailing sheath of the first bubble as well as the leading sheath of the next bubble. This raises the strength of the sheath fields and increases the number of bubbles that fit into this snapshot at $r_L = 33$, consistent with the greater power produced by this sharpened current profile. Referring to Eq. (11.3), the increase of field strength corresponds to a doubling of the factor in brackets from 32 to 64, and the increase in the number of bubbles from 5 to 7 corresponds to a decrease of δ in the denominator from 7 to 5. When $\Delta r \gg 2$, the latter decrease becomes small so that the main increase in power is due to the constructive interference of the fields in the overlapping sheaths.

Next, we extend this analysis to non-linear ramps, like the sinusoidal profile whose radiation is already familiar to us from more conventional treatments. The middle portion of a non-linear ramp provides an important contribution to the radiated energy and cannot be neglected as it could for the linear current profile. Consequently, the energy radiated during the half-period, δ, as given in Eq. (11.2), now has an extra term:

$$\frac{\mathcal{E}}{\mathcal{E}_0} = -\frac{1}{r^2}\left[\int_{r+\Delta r}^{r+\delta} \frac{\partial F_{2o}}{\partial r}\frac{\partial F_{2o}}{\partial r_L}\,dr_L + \int_{r+\delta}^{r+\Delta r+2} \frac{\partial F_3}{\partial r}\frac{\partial F_3}{\partial r_L}\,dr_L \right.$$
$$\left. + \int_{r+\Delta r+2}^{r+\delta+2} \frac{\partial F_{1o}}{\partial r}\frac{\partial F_{1o}}{\partial r_L}\,dr_L + \int_{r+\delta+2}^{r+\delta+\Delta r} \frac{\partial F_{mo}}{\partial r_L}\frac{\partial F_{mo}}{\partial r}\,dr_L\right], \qquad (11.4)$$

where $F_{1o} = F_{1o}(r, r_L - \delta)$ refers to the leading sheath of the off-bubble, $F_{2o} = F_{2o}(r, r_L)$ refers to the trailing sheath of the on-bubble, $F_3 = F_{2o}(r, r_L) - F_{1o}(r, r_L - \delta)$ refers to their overlap, and $F_{mo} = F_{mo}(r, r_L - \delta)$ refers to the middle region of the on-bubble.

For a continuous extension between the end of the on-ramp and the start of the off-ramp, we set $\delta = \Delta r$ in Eq. (11.1). Then, making a change of variables in the F_{mo}-dependent integral, we obtain

$$\frac{\mathcal{E}}{\mathcal{E}_0} = -\frac{1}{r^2}\left[\int_{r+\Delta r}^{r+\Delta r+2} \frac{\partial F_3}{\partial r}\frac{\partial F_3}{\partial r_L}\,dr_L + \int_{r+2}^{r+\Delta r} \frac{\partial F_{mo}}{\partial r}\frac{\partial F_{mo}}{\partial r_L}\,dr_L\right], \qquad (11.5)$$

where $F_{mo} = F_{mo}(r, r_L - \Delta r)$ refers to the middle of the on-bubble, and $F_3 = F_{2o}(r, r_L) - F_{1o}(r, r_L - \Delta r)$ refers to the overlap between the trailing sheath of the on-bubble and the leading sheath of the off-bubble.

For $\Delta r \gg 2$, we can evaluate these two terms in the same way that we did for the non-linear ramp in Chap. 10. The second term in Eq. (11.5) reduces to the $(\Delta)^{-3}$-dependent expression in Eqs. (10.16) and (10.42). If the ramp is curved at both ends, the first term in Eq. (11.5) reduces to a $(\Delta)^{-4}$-dependent expression that can be neglected when $\Delta r \gg 2$. Thus, for a smooth, sinusoidal-like variation, the details of the on–off connection are unimportant, and the constructive interference gives an energy that is small compared to the dominant contribution from the middle of the ramp.

Let us illustrate this result using the sinusoidal current profile $G(\rho) = \sin^2\{(\pi/2)(\rho/\Delta r)\}$, where ρ lies in the range $(0, \Delta r)$. In this case, the current is always positive, oscillating between 0 and 1 about an average value of 1/2. After obtaining the F-values from Eq. (5.7) with $dr'_L/\Delta r$ replaced by $G'(r_L - r'_L)dr'_L$, we then obtain $\mathcal{E}_{mo}/\mathcal{E}_0 = (\pi^4/8)(\Delta r)^{-3} - (\pi^4/2)(\Delta r)^{-4} + \mathcal{O}(\Delta r)^{-5}$ from the F_{mo}-term in Eq. (11.5) and $\mathcal{E}_3/\mathcal{E}_0 = +(\pi^4/2)(\Delta r)^{-4} - (4\pi^4/15)(\Delta r)^{-6} + \mathcal{O}(\Delta r)^{-8}$ from the F_3-term (i.e., the interference term). Thus, for $\Delta r \gg 2$, we can neglect the interference. However, if we did not neglect the interference, its $(\Delta r)^{-4}$-dependence would be cancelled out by the $(\Delta r)^{-4}$-dependence of the F_{mo} contribution, and make the sum even closer to the $(\Delta r)^{-3}$-dependent term obtained from Eqs. (10.16) or (10.42). Dividing this energy by Δr, we obtain the radiated power $p/p_0 = (1/8)(\pi/\Delta r)^4$ when $\Delta r \gg 2$. Because $\pi/\Delta r = \omega$, the frequency of the oscillation, it follows that $p/p_0 = \omega^4/8$ in this low-frequency regime.

In this example, the current repeatedly turned on and off, but never changed sign. Let us see what happens when we allow the current to change sign. We can obtain such a current by doubling the strength of this current and then subtracting 1. In this case, the new current profile is $g(\rho) = 2G(\rho) - 1 = -\cos(\pi\rho/\Delta r)$, which oscillates between -1 and $+1$ with the same frequency $\omega = \pi/\Delta r$. Substituting this value of g into Eq. (10.17) gives $\mathcal{E}/\mathcal{E}_0 = (\pi^4/2)(\Delta r)^{-3}$ and $p/p_0 = (1/2)(\pi/\Delta r)^4$, which is 4 times the energy and power obtained from the current that varies between 0 and 1. Thus, the radiated power depends on the square of the current amplitude as well as the fourth power of the frequency, giving a general relation $p/p_0 = (1/2)\omega^4 A^2$, where A is the center-to-peak amplitude of the oscillation. This result can be obtained by substituting $G(\rho) = A\cos(\omega\rho)$ into Eq. (10.17), and it is a well-known formula for the radiation from an oscillating dipole of amplitude, A, and angular frequency, ω (Slater and Frank 1947).

We have seen that a periodic extension of a linear ramp gave twice as much radiated power when the turn-off occurred promptly at the time Δr than when it was delayed by the time, $2R/c$. This extra power occurred because the field in the trailing sheath of the first bubble interfered constructively with the field in the leading sheath of the second bubble, and because the radiated energy occurred in these sheaths and not in the central region between them. If the radiated energy were not confined to these sheaths, but were distributed broadly across the bubble, then the overlap would not have mattered, as we found for the sinusoidal profile. Next, we will find that when the repetition time is less than $2R/c$, the interference will affect the radiated power.

11.2 Overlap with $\Delta r < 2$

Consider what happens in the "short wavelength limit" when the current is repeatedly switched on and off. To illustrate this case, we suppose that the current is changed suddenly and repeatedly at intervals of τ (in units of R/c) to produce a continuous "square-wave" profile. This current creates a continual stream of magnetic bubbles moving outward away from the sphere. The radiated energy per unit time is the individual bubble energy, $3\mathcal{E}_0/2$, divided by the time, τ, between switchings. Thus, the average radiated power, p, is

$$p/p_0 = \frac{3}{2}\frac{1}{\tau}, \quad \tau \geq 2, \tag{11.6}$$

where $p_0 = \mathcal{E}_0 c/R$. From Eq. (11.6), we see that the shorter the interval between switchings is, the less wasted time there will be and the greater will be the average radiated power. When $\tau = 2$, no space is left between consecutive bubbles, and $p/p_0 = 3/4$. However, as τ becomes less than 2, the bubble produced by turning the current off will start to come out of the sphere before the first bubble has left. Consequently, these two radiation fields will interfere and cause the radiated energy to differ considerably from the sum of the energies of the individual bubbles.

The result depends on whether the interference is constructive or destructive. As τ falls below 2, the trailing part of one bubble overlaps with the leading part of the next, and these like-polarity fields interfere constructively. This constructive interference increases until τ reaches 1 and the radiated power reaches a maximum. As τ falls below 1, destructive interference begins to occur, causing the energy to decrease again until τ reaches the next turning point at 2/3.

To calculate the radiated power, we recognize that the "square-wave" current profile of "width," τ, corresponds to an F-value that is the superposition of terms:

$$F(r, r_L) = \sum_{k=0}^{\infty} (-1)^k \, f_o(r, r_L + k\tau). \tag{11.7}$$

The idea here is to use this F-value to evaluate the radiated energy during an interval $(r, r + \tau)$ and then divide that energy by the duration, τ, to obtain the average radiated power. In this case, we use Eq. (9.15) applied to the interval $(r, r + \tau)$

$$\frac{\mathcal{E}_{rad}}{\mathcal{E}_0} = -\frac{1}{r^2} \int_r^{r+\tau} \frac{\partial F}{\partial r} \frac{\partial F}{\partial r_L} \, dr_L. \tag{11.8}$$

Next, we need to determine how many terms of the sum contribute to the energy from that interval. As we consider progressively larger values of k, we eventually find that $r_L + k\tau \geq r + 2$. In that case, $f_o(r, r_L + k\tau) = 1$, $\partial f_o/\partial r = 0$ and this term does not contribute to the radiated energy from the interval $(r, r + \tau)$. When $r_L = r$, this condition reduces to $k\tau \geq 2$, and the critical value of k is $2/\tau$. In general, $2/\tau$ is not an integer, so we select its integral part $m = \text{int}(2/\tau)$. Now, for $r_L \geq r + 2 - m\tau$, $f_o(r, r_L + m\tau)$ does not contribute to the radiated energy from the interval $(r, r + \tau)$, but for $r_L < r + 2 - m\tau$, it does contribute. Consequently, we break the interval into two parts and write

$$\frac{\mathcal{E}_{rad}}{\mathcal{E}_0} = -\frac{1}{r^2} \int_r^{r+2-m\tau} \frac{\partial F}{\partial r} \frac{\partial F}{\partial r_L} \, dr_L - \frac{1}{r^2} \int_{r+2-m\tau}^{r+\tau} \frac{\partial F_{red}}{\partial r} \frac{\partial F_{red}}{\partial r_L} \, dr_L, \tag{11.9}$$

where F includes the term with $k = m$

$$F(r, r_L) = \sum_{k=0}^{m} (-1)^k \, f_o(r, r_L + k\tau), \tag{11.10}$$

and F_{red} is the reduced value of F with the term $k = m$ omitted

$$F_{red}(r, r_L) = \sum_{k=0}^{m-1} (-1)^k \, f_o(r, r_L + k\tau). \tag{11.11}$$

Taking f_o from Eq. (2.11a) and performing the integration, we obtain

$$\frac{\mathcal{E}}{\mathcal{E}_0} = \begin{cases} +\frac{3}{16} \left[8 - 12m\,\tau + m^2\,(2m + 3)\,\tau^3 \right], & m \text{ even} \\ -\frac{3}{16} \left[8 - 12(m + 1)\,\tau + (m + 1)^2\,(2m - 1)\,\tau^3 \right], & m \text{ odd.} \end{cases} \tag{11.12}$$

Dividing these energies by τ, we obtain the average radiated power as a function of the time difference, τ, between the sudden switch-on's and switch-off's of the current.

$$\frac{p}{p_0} = \begin{cases} +\frac{3}{16}\left[8/\tau - 12m + m^2\,(2m+3)\,\tau^2\right], & m \text{ even} \\ -\frac{3}{16}\left[8/\tau - 12(m+1) + (m+1)^2\,(2m-1)\,\tau^2\right], & m \text{ odd.} \end{cases} \tag{11.13}$$

Although these expressions were obtained for $\tau \leq 2$, when $m = 0$ they give $p/p_0 = 3/2\tau$, in agreement with Eq. (11.6). Because Eq. (11.6) applies for $\tau \geq 2$, this means that Eqs. (11.12) and (11.13) are valid for all τ in the range $(0, \infty)$. To examine the behavior at the small-τ end of the spectrum, it is convenient to express the power in terms of a "frequency" $\nu = 2/\tau$. In this case, Eq. (11.13) becomes

$$\frac{p}{p_0} = \begin{cases} +(3/4\nu^2)\left[(\nu - m)^3 - 3m^2\,\{\nu - (m+1)\}\right], & m \text{ even} \\ -(3/4\nu^2)\left[-3(m+1)^2\,(\nu - m) + \{\nu - (m+1)\}^3\right], & m \text{ odd,} \end{cases} \tag{11.14}$$

where $m = \text{int}(2/\tau)$ represents the integral value of ν. In these equations, the first terms come from the F-dependent integral in Eq. (11.9) and the second terms come from the F_{red}-dependent integral.

Figure 11.2 shows this radiated power as a function of $\nu = 2/\tau$. We see the expected linear increase from 0 to 3/4 as ν increases from 0 to 1 while the non-interfering bubbles crowd closer together. As ν exceeds 1, the bubbles start to interfere constructively and the power increases rapidly to 9/4 as ν increases to 2 where the overlap is greatest. This maximum power is 3 times the power radiated when $\nu = 1$, $\tau = 2$, and the bubbles are tangent. As ν increases further, destructive interference sets in and the power falls to a local minimum at $\nu = 3$. Thereafter, the power oscillates between a maximum value of 9/4 when ν is an even integer and

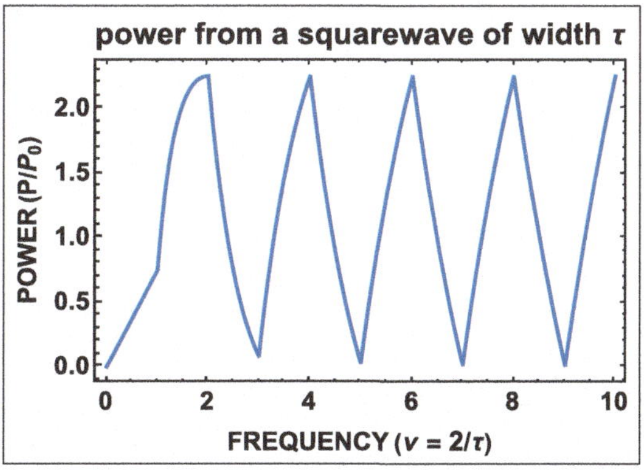

Fig. 11.2 Average radiated power from a square-wave current ramp of width τ, plotted as a function of the frequency $\nu = 2/\tau$

a minimum value of $(3/4)\nu^{-2}$ when ν is an odd integer, as one can see by setting $\nu = m$ in each part of Eq. (11.14). Also, because the first terms of Eq. (11.1) vanish when $\nu = m$, the extreme values of p/p_0 must come from the F_{red}-dependent integral in Eq. (11.9).

Thus, if we regard the sphere as an antenna, then it will radiate most efficiently if its "frequency" $\nu = 2n$ or its pulse duration $\tau = 1/n$, where $n = 1, 2, 3, \ldots$. Expressed in terms of the "wavelength" $\lambda = 2\tau R = (4/\nu)R$, this condition is $R = n\lambda/2$. Thus, our spherical antenna will radiate most efficiently if its radius is an integral number of half-waves. This illustrates the general principle that a "half-wave" antenna ($R = \lambda/2$) is the shortest one for which there is full constructive interference between the consecutively emitted bubbles, corresponding to the maximum average radiated power.

Figure 11.3 shows snapshots of the field lines in an equatorial region located 25–30R from the sphere. These plots were calculated for square-wave current profiles with $\nu = 1, 2, 3$, and 4. When $\nu = 1$, the bubbles just touch along the dotted lines where the trailing polarity of one bubble matches the leading polarity of the next bubble (despite their red and blue colors which indicate clockwise and anti-clockwise circulation, not magnetic polarity). The width of each bubble is 2 units of R/c, corresponding to the separation of the dotted lines. When $\nu = 2$, the bubbles overlap constructively, strengthening the leading and trailing fields and shrinking the widths to 1 unit. When $\nu = 3$, the fields interfere destructively, weakening the field so much that only the dashed boundaries are visible. Finally, when $\nu = 4$, the fields interfere constructively again and their widths have shrunk to 1/2 of a unit of R/c.

Figure 11.4 shows the dependence of the field-line patterns on the amount of overlap, τ, and on the character of the profile. The field lines in the left panels were generated from a square-wave current profile that ranges between 0 and 1, whereas the field lines in the right panels were generated from a square-wave profile ranging from -1 to $+1$. By doubling the amplitude of F, we might expect to quadruple the radiated energy, given by Eq. (11.8). Also, because the square wave with alternating signs is symmetric about zero, we would expect its consecutive bubbles to have the same width, as shown in the lower right panel of Fig. 11.4. On the other hand, because the square wave that alternates between 0 and 1 is symmetric about $+1/2$, rather than 0, we might expect its clockwise (red) bubbles to be larger than its counterclockwise (blue) bubbles. Moreover, their average width should match the widths of the bubbles produced by the signed profile, which are 1 unit of R/c when $\tau = 1$. On the other hand, if there is no overlap between consecutive bubbles, as in the upper panels where $\tau = 2$ and $\nu = 1$, we would expect all of the bubbles to have a common width of 2 units of R/c.

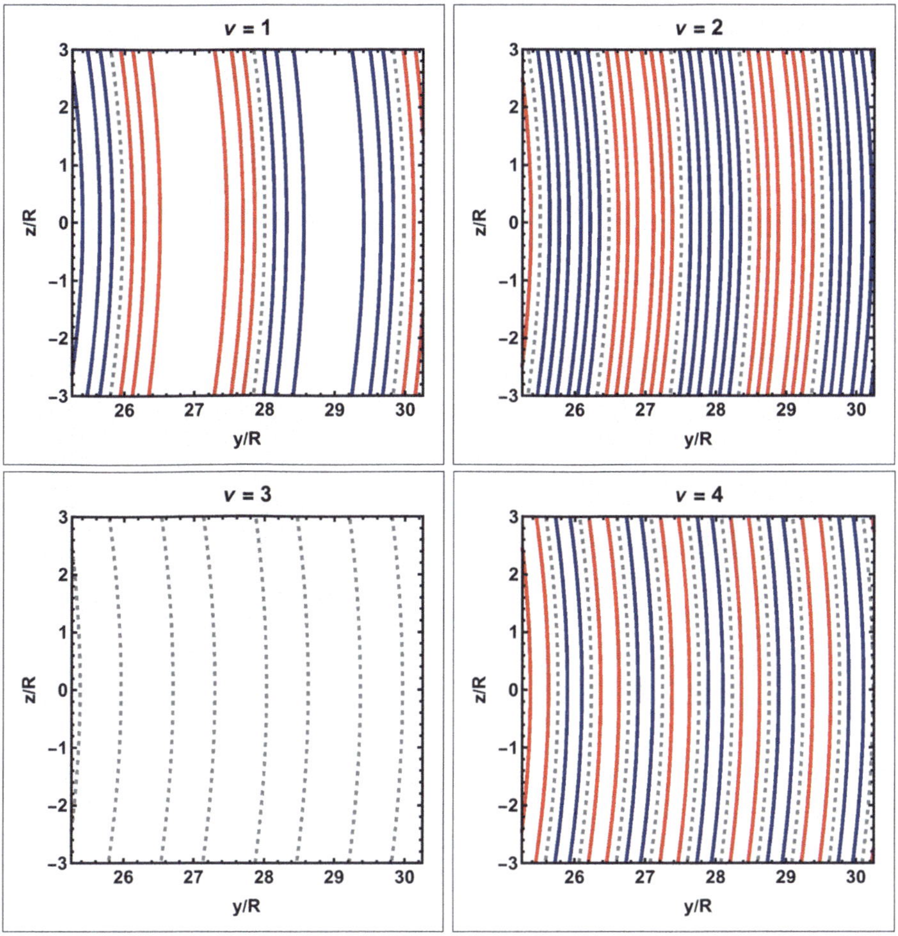

Fig. 11.3 Near-equatorial field lines at radial distances in the range 25–30 R, produced by a square-wave current profile whose "frequency" $\nu = 1, 2, 3,$ and 4 c/R. (Upper left) $\nu = 1$, showing the contiguous red and blue bubbles of clockwise and counterclockwise directed flux, respectively. (Upper right) $\nu = 2$, showing the constructive superposition caused by fully overlapping bubbles. (Lower left) $\nu = 3$, showing the absence of fields produced by the fully destructive overlap. (Lower right) $\nu = 4$, showing the fields obtained at the second peak of constructive overlap. The dashed lines mark the boundaries between consecutive bubbles

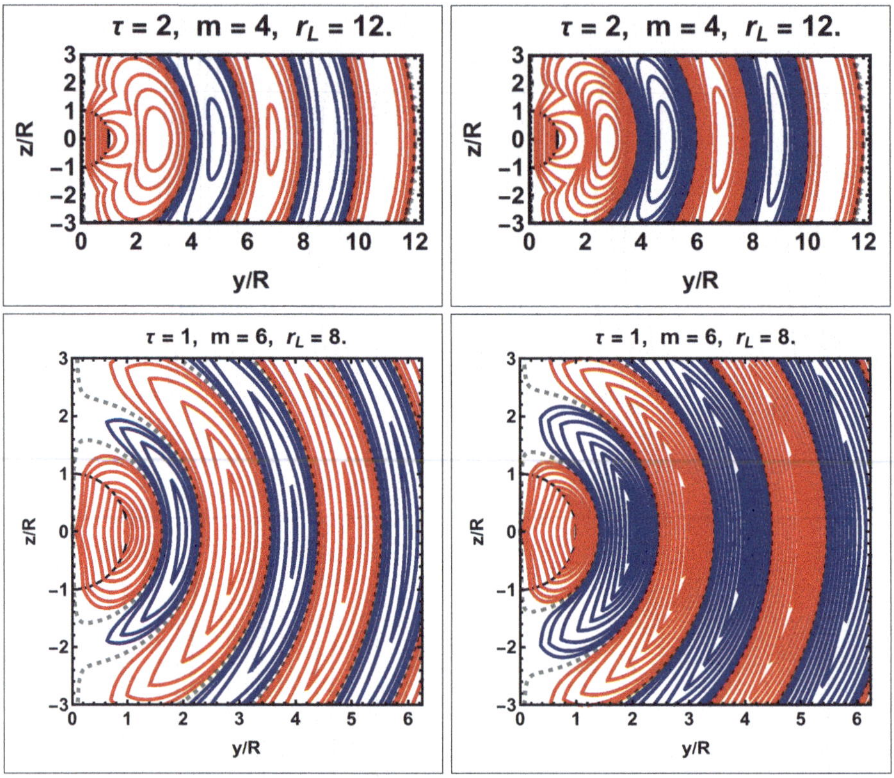

Fig. 11.4 Magnetic field lines for "square-wave" current profiles with repetition times $\tau = 2$, corresponding to adjacent bubbles (top), and $\tau = 1$, corresponding to bubbles with fully constructive overlap (bottom). (Left panels) The current jumps from 0 to 1 and back, corresponding to an oscillation about $+1/2$ with an amplitude of $\pm 1/2$. (Right panels) After the first increase to $+1$, the current jumps to -1 and back to $+1$, corresponding to an oscillation about 0 with an amplitude of ± 1. This larger-amplitude profile produces 4 times the power of the smaller-amplitude profile. m indicates the number of current changes after the initial turn-on with $m = 0$. The dashed lines mark the boundaries between consecutive bubbles

Chapter 12
Single-Source Dynamics

12.1 Force

We have seen that the sudden turn-on of the spherical current produces an outward-moving bubble whose wake is the static dipole field of a uniformly magnetized sphere. Not only does this process impart energy to the bubble and the static field, but also it imparts momentum to the bubble and exerts a force on the current.

We can analyze these effects using the well-known relation

$$F_i = \int_{\text{box}} T_{ij} dA_j - \frac{\partial p_i}{\partial t}, \tag{12.1}$$

which is the volume integral of

$$\mathbf{J}\times\mathbf{B} = \frac{(\nabla\times\mathbf{B})\times\mathbf{B}}{\mu_0} + \epsilon_0(\nabla\times\mathbf{E})\times\mathbf{E} - \frac{\partial}{\partial t}(\epsilon_0\mathbf{E}\times\mathbf{B}), \tag{12.2}$$

written in tensor notation. Here, F_i is the ith component of the net force on currents within the closed surface indicated by the word "box," T_{ij} is the Maxwell stress tensor given by

$$T_{ij} = B_i B_j/\mu_0 - \delta_{ij} B^2/2\mu_0 + \epsilon_0 E_i E_j - \delta_{ij}\epsilon_0 E^2/2, \tag{12.3}$$

and p_i is the ith component of the field momentum given by

$$p_i = \frac{1}{c^2}\int_V \left(\frac{\mathbf{E}\times\mathbf{B}}{\mu_0}\right)\cdot \mathbf{e_i} dV = \frac{1}{c^2}\int_V S_i dV, \tag{12.4}$$

where V indicates the volume enclosed by the "box" and S_i is the ith component of the Poynting flux, $\mathbf{S} = \mathbf{E}\times\mathbf{B}/\mu_0$.

N. R. Sheeley, Jr., *Transient Magnetic Fields*,
https://doi.org/10.1007/978-3-030-40264-8_12

If the box consists of a sphere concentric with the current source, then the symmetry causes the net stress, momentum, and force to vanish. However, if the symmetry is broken by offsetting the box from the center of the current source, then these quantities will not vanish when averaged over the box. In this subsection, our objective is to determine the force, F_y, on the current as a function of the time r_L.

For this purpose, it is convenient to think of this relation in terms of the stress on a closed surface. This stress gives the net force on any electric currents that may lie in the volume enclosed by the surface plus the time rate of change of the field momentum in that volume. Thus, we can determine the force on an electric current by selecting a surface that isolates the current and also provides a geometry convenient for evaluating the stress. Likewise, we can determine the field momentum by choosing a momentum-isolating surface.

We begin by considering a tight-fitting glove around the $(y > 0)$-portion of the spherical current. Such a region would be obtained by letting the radii of the dashed circles in the left panel of Fig. 12.1 approach the radius of the sphere. In this case, the volume integral for p_y in Eq. (12.4) would vanish, and the force, F_i, in Eq. (12.1) would be given by the surface integral alone. For this surface integral, there is no contribution from the vanishingly small end face at $y = 0$, so that only the oppositely directed contributions from the virtually coincident outer and inner spherical surfaces will contribute. The discontinuity of the stress tensor, associated with the presence of the current and the jump of B_θ, prevents these opposing contributions from cancelling out and making the force vanish.

The net force in the y-direction can be expressed in terms of the components, dF_r and dF_θ, on a patch of area dA_r. Using the components of the stress tensor

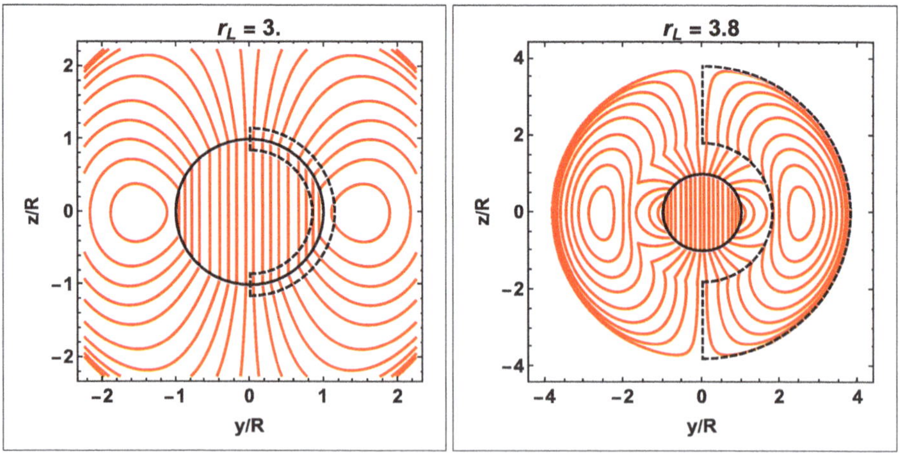

Fig. 12.1 Dashed lines, showing the boxes used to calculate the y-component of the stress exerted on the spherical current (left) and on the outgoing bubble (right) by the electromagnetic field

$$T_{rr} = \frac{B_r^2 - B_\theta^2}{2\mu_0} - \frac{1}{2}\epsilon_0 E_\phi^2 \tag{12.5a}$$

$$T_{\theta r} = \frac{B_\theta B_r}{\mu_0}, \tag{12.5b}$$

and recognizing that only B_θ is discontinuous, we can write

$$dF_r = \Delta T_{rr}\, dA_r = -\Delta\left(\frac{B_\theta^2}{2\mu_0}\right) dA_r \tag{12.6a}$$

$$dF_\theta = \Delta T_{\theta r}\, dA_r = \frac{B_r \Delta B_\theta}{\mu_0}\, dA_r, \tag{12.6b}$$

where the prefix Δ refers to the jump as one moves outward across the surface. Next, we extract the θ-dependence by introducing θ-less field components b_r, b_θ, and $\mathcal{E}_\phi$:

$$B_r/B_0 = b_r(r, r_L)\ \cos\theta, \tag{12.7a}$$

$$B_\theta/B_0 = b_\theta(r, r_L)\ \sin\theta, \tag{12.7b}$$

$$E_\phi/cB_0 = \mathcal{E}_\phi(r, r_L)\ \sin\theta. \tag{12.7c}$$

The dependence of the components on r and r_L is given by

$$b_r(r, r_L) = r^{-3}\ f(r, r_L), \tag{12.8a}$$

$$b_\theta(r, r_L) = (1/2)\ r^{-3}\left[f(r, r_L) - r\ \partial f(r, r_L)/\partial r\right], \tag{12.8b}$$

$$\mathcal{E}_\phi(r, r_L) = -(1/2)\ r^{-2}\ \partial f(r, r_L)/\partial r_L, \tag{12.8c}$$

with $f(r, r_L)$ defined in Eq. (2.10). Also, $f = f_o$ outside the sphere and $f = f_i$ inside the sphere, as given in Eqs. (2.11a) and (2.11b)

With these definitions, the force components on the surface at $r = 1$ become

$$dF_r = -\sin^2\theta\ \Delta\left(b_\theta^2\right) dA_r \left(\frac{B_0^2}{2\mu_0}\right), \tag{12.9a}$$

$$dF_\theta = +2\sin\theta\cos\theta\ b_r \Delta b_\theta\ dA_r \left(\frac{B_0^2}{2\mu_0}\right). \tag{12.9b}$$

For r_L in the range (1, 3), Eqs. (12.8) may be used to obtain b_r, Δb_θ, and $\Delta(b_\theta^2)$. Setting $r = 1$ we find that $b_r = f(1, r_L) = (r_L - 1)[6 - (r_L - 1)^2]/4$, $\Delta b_\theta = 3/2$, and $\Delta b_\theta^2 = -(3/8)(r_L - 1)^3$. In this case, the time-varying, differential force components become

$$dF_r = \frac{3}{8}\,\sin^2\theta\,(r_L - 1)^3\left(\frac{B_0^2}{2\mu_0}\right) dA_r, \tag{12.10a}$$

$$dF_\theta = \frac{3}{4}\,\sin\theta\cos\theta\,(r_L - 1)\left[6 - (r_L - 1)^2\right]\left(\frac{B_0^2}{2\mu_0}\right) dA_r. \tag{12.10b}$$

When $r_L > 3$, we must use the static values $f_i = r^3$ for the uniform field left inside the sphere and $f_o = 1$ for the dipole field left outside, as discussed previously. Substituting these values of f into Eq. (12.8), we obtain $b_\theta = -1$ inside and $b_\theta = (1/2)r^{-3}$ outside. Then, setting $r = 1$ we find that $b_r = 1$, $\Delta b_\theta = 3/2$, and $\Delta b_\theta^2 = -3/4$, giving the static force components for the uniformly magnetized sphere that is left behind the outgoing bubble

$$dF_r = \frac{3}{4}\,\sin^2\theta\left(\frac{B_0^2}{2\mu_0}\right) dA_r, \tag{12.11a}$$

$$dF_\theta = 3\,\sin\theta\cos\theta\left(\frac{B_0^2}{2\mu_0}\right) dA_r. \tag{12.11b}$$

At this point, it is interesting to note that the radial components are always directed outward and the poloidal components are always directed toward the equator. Thus, if the individual current elements were not fixed on the sphere by external mechanical forces (like staples or glue), they would slide toward the equator and merge into a single expanding loop.

Finally, we resolve these forces into the y-direction using the relation $dF_y = (dF_r\sin\theta + dF_\theta\cos\theta)\sin\phi$, and then integrate over the hemisphere with $y > 0$ and $r = 1$ using the areal element $dA_r = \sin\theta d\theta d\phi$. The result is

$$F_y/F_0 = \begin{cases} (3/32)(r_L - 1)[(r_L - 1)^2 + 12], & 1 < r_L < 3, \\ 21/16 \approx 1.3, & r_L > 3, \end{cases} \tag{12.12}$$

where $F_0 = (\pi R^2)(B_0^2/2\mu_0)$. This function is plotted in Fig. 12.2.

Here, we can see that the force, F_y, increases monotonically from 0 to $3F_0$ as the transient moves inward and then outward through the sphere. Based on Eq. (12.12), the initial increase is approximately linear, varying as $F_y/F_0 \approx (9/8)(r_L - 1)$ while $r_L - 1 \ll \sqrt{12} \approx 3.5$ as the interior wave approaches the center of the sphere. The force increases slightly faster after the wave reaches the center and heads outward. When the wave eventually crosses the surface and departs from the sphere, F_y drops discontinuously to the static value, $(21/16)F_0$. This static value is also the average impulse, $(1/2)\int_1^3 F_y dr_L$, of the wave while it encountered the current during the interval $(1, 3)$.

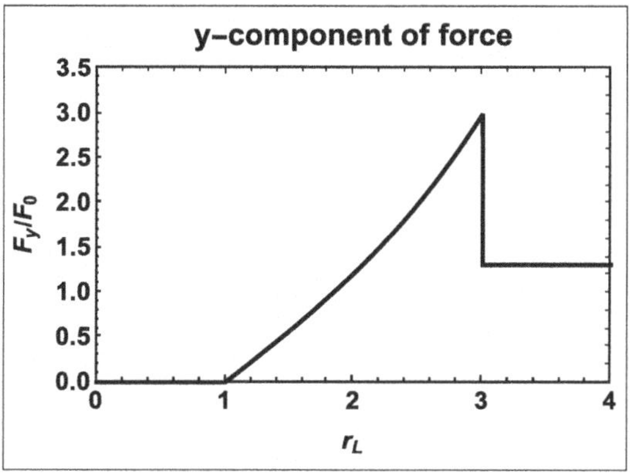

Fig. 12.2 The y-component of force plotted versus time, r_L. The force increases gradually to its maximum value of $F_y/F_0 = 3$ when $r_L = 3$ and drops suddenly to $F_y/F_0 = 21/16 \approx 1.3$ when the bubble leaves the sphere. The unit of force is $F_0 = (B_0^2/2\mu_0)(\pi R^2)$

12.2 Momentum

As we mentioned above, the symmetry properties of the fields cause the net momentum to vanish in a region centered on the current source. However, when integrating over part of the region, such as the hemisphere where $y > 0$, the momentum does not vanish, and our next objective is to determine its value.

We refer to the right panel of Fig. 12.1 whose dashed line represents the boundary of the volume, V, used in evaluating the integral in Eq. (12.4). Of course, this dashed line indicates the boundary when the bubble lies fully outside the sphere at $r = 1$. More generally, we can combine hemispherical surfaces at the leading and trailing edge with the annular face that joins them at $y = 0$, even when the trailing hemisphere lies inside the sphere. The idea is to use the field components given in Eq. (12.7) to evaluate $\mathbf{E}\times\mathbf{B}$ and its y-component $(\mathbf{E}\times\mathbf{B})_r \sin\theta \sin\phi + (\mathbf{E}\times\mathbf{B})_\theta \cos\theta \sin\phi$. Multiplying this expression by the volume element $dV = r^2 \sin\theta d\theta d\phi dr$, and then integrating over r in the range (r_F, r_L), and the angles θ and ϕ in the range $(0, \pi)$, we obtain the y-component of momentum, p_y

$$\frac{p_y}{p_0} = \left(\frac{1}{2}\right) \int_{r_F}^{r_L} \mathcal{E}_\phi \,(b_r - 3b_\theta)\, r^2 dr, \tag{12.13}$$

where $p_0 = F_0 R/c$ is the unit of momentum and $F_0 = (\pi R^2)(B_0^2/2\mu_0)$. The lower limit, r_F, depends on whether the wave inside the sphere is headed inward or outward. In particular, $r_F = 2 - r_L$ if $r_L < 2$, and $r_F = r_L - 2$ if $r_L > 2$. Also, f and its derived quantities b_r and b_θ depend on whether the contribution originates inside or outside the sphere. When $r_L < 3$, part of the transient lies inside the sphere where the plus sign must be used in the definition of f and part of the transient lies outside the sphere where the minus sign must be used. The function, $\mathcal{E}_\phi$, has the same value regardless of the sign used in the definition of f. When $r_L > 3$, the transient has left the sphere, so only the exterior values need to be used.

Another way of saying this is that inside the sphere the momentum, p_y^{in}, is given by

$$\frac{p_y^{in}}{p_0} = \begin{cases} \left(\frac{1}{2}\right) \int_{2-r_L}^{1} \mathcal{E}_\phi (b_r - 3b_\theta)\, r^2 dr, \; 1 < r_L < 2, \\ \left(\frac{1}{2}\right) \int_{r_L-2}^{1} \mathcal{E}_\phi (b_r - 3b_\theta)\, r^2 dr, \; 2 < r_L < 3, \\ 0, \qquad r_L > 3, \end{cases} \tag{12.14}$$

where b_r and b_θ are evaluated inside the sphere using $f = f_i$. Likewise, p_y^{out} is given by

$$\frac{p_y^{out}}{p_0} = \begin{cases} \left(\frac{1}{2}\right) \int_{1}^{r_L} \mathcal{E}_\phi (b_r - 3b_\theta)\, r^2 dr, \quad 1 < r_L < 2, \\ \left(\frac{1}{2}\right) \int_{1}^{r_L} \mathcal{E}_\phi (b_r - 3b_\theta)\, r^2 dr, \quad 2 < r_L < 3, \\ \left(\frac{1}{2}\right) \int_{r_L-2}^{r_L} \mathcal{E}_\phi (b_r - 3b_\theta)\, r^2 dr, \; r_L > 3, \end{cases} \tag{12.15}$$

where b_r and b_θ are evaluated outside the sphere using $f = f_o$. The total momentum, p_y^{tot}, at the time, r_L, is then obtained by adding p_y^{in} and p_y^{out} evaluated at this time.

Although these integrals can be evaluated analytically as a function of r_L, the resulting expressions for p_y are complicated series of both logarithmic and algebraic terms. For example, when $r_L > 3$, p_y^{out}/p_0 is given by the sum of the two terms

$$a(r_L) = -(1/32)\left(21r_L^3 - 84r_L^2 + 76r_L + 19\right)/(r_L - 2) \tag{12.16a}$$

$$b(r_L) = -(3/64)\, r_L \left(7r_L^2 - 21r_L + 15\right) ln(1 - 2/r_L), \tag{12.16b}$$

whose algebraic forms do not provide much insight into the evolution of the momentum. So, rather than reproducing all of the algebraic results here, we instead plot the three terms p_y^{in}, p_y^{out}, and p_y^{tot} in order to gain some insight into how the momentum evolves. Figure 12.3 shows the result. As one can see from the dashed curve, the interior momentum is initially negative as the ingoing wave moves toward the center of the sphere. This momentum reaches a minimum value of about -0.55 at $r_L = 1.75$ before increasing again. However, the momentum remains negative even when the wave reaches the center, and it does not become positive until $r_L = 2.15$. This interior momentum reaches a positive maximum of about 0.45 when $r_L = 2.6$. In contrast, the exterior momentum is always positive and has much less variation, as shown by the red curve. Initially, the total momentum becomes negative as the interior contribution outweighs the exterior one. The momentum remains negative for a while, reaching its minimum value of about -0.17 at $r_L = 1.73$, before turning around and becoming positive at $r_L = 1.95$, just before the ingoing wave reaches the center of the sphere. Then the total momentum increases to a maximum value of about 0.8 near $r_L = 2.6$ before falling again to a value slightly

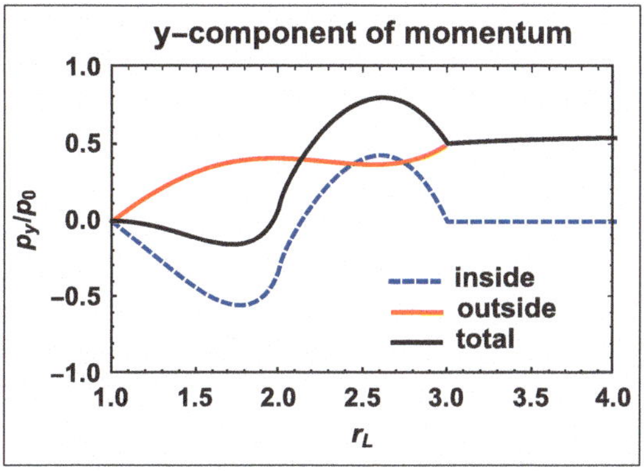

Fig. 12.3 The y-component of field momentum plotted versus time, r_L. The red line shows the contribution from outside the sphere; the dashed line is from inside the sphere; and the black line shows the total momentum from inside and outside the sphere. The unit of momentum is $p_0 = F_0 R/c = (\pi R^2)(B_0^2/2\mu_0)(R/c)$

more than 0.5 at $r_L = 3$ when the bubble leaves the sphere. This alternating variation of momentum approximately matches the inward and outward motion of the trailing wave.

We can gain further insight by calculating the momentum approximately and expanding the results about $r_L = 1$ and $r_L = 3$. (The momentum has a vertical slope at $r_L = 2$, so the power series about $r_L = 2$ diverges.) First, consider the initial behavior for r_L in the range (1, 2). In this case, we can use Eq. (12.13) with $r_F = 2 - r_L$ to obtain the momentum, p_y. We take the interior and exterior values of b_r, b_θ, and $\mathcal{E}_\phi$ in the subintervals $(2 - r_L, 1)$ and $(1, r_L)$, respectively (which is equivalent to adding p_y^{in} and p_y^{out} for r_L in the range (1, 2), as given in Eqs. (12.14) and (12.15)). Expanding the result in powers of $(r_L - 1)$, we find that

$$\frac{p_y}{p_0} = -\frac{9}{16}(r_L - 1)^2 + \frac{57}{128}(r_L - 1)^4 + \ldots \tag{12.17}$$

with $p_0 = F_0 R/c$. Setting $R = c = 1$ and differentiating this equation with respect to r_L, we obtain

$$\frac{1}{F_0}\frac{\partial p_y}{\partial r_L} = -\frac{9}{8}(r_L - 1) + \frac{57}{32}(r_L - 1)^3 + \ldots . \tag{12.18}$$

By comparison, the force exerted on the current is given exactly by Eq. (12.12) as

$$\frac{F_y}{F_0} = \frac{9}{8}(r_L - 1) + \frac{3}{32}(r_L - 1)^3, \tag{12.19}$$

where $F_0 = (\pi R^2)(B_0^2/2\mu_0)$. Thus, to first order, the force on the electric current equals the initial rate of decrease of field momentum.

This means that the net stress on a closed surface surrounding the $(y > 0)$-portion of the disturbance must initially vanish to first order. It is easy to confirm this conclusion by using a surface that consists of hemispheres at radii slightly greater than r_L and slightly less than $2 - r_L$ joined by the annular end face at $y = 0$. There is no contribution at these hemispheres because they lie in field-free regions. The contribution at the annular end face where $y = 0$ may be obtained using the T_{yy} component of the field tensor, which reduces to

$$T_{yy} = -\left(\frac{B_r^2 + B_\theta^2}{2\mu_0}\right) + \frac{1}{2}\epsilon_0 E_\phi^2 \tag{12.20}$$

because $\phi = 0$ and therefore $B_y^2 = 0$ and $E_y^2 = E_\phi^2$. In this case, we obtain

$$\frac{1}{F_0}\int T_{yy} dA_y = +\int_{2-r_L}^{r_L} \left(b_r^2 + b_\theta^2 - \mathcal{E}_\phi^2\right) r dr, \tag{12.21}$$

with the interior and exterior values of b_r, b_θ, and $\mathcal{E}_\phi$ obtained for the regions $(2 - r_L, 1)$ and $(1, r_L)$, respectively. The positive sign has been chosen to allow for the fact that the y-axis points in the direction of the inner normal rather than the outer normal required in Eq. (12.1). In this case, we obtain

$$\frac{1}{F_0}\int T_{yy} dA_y = \frac{15}{8}(r_L - 1)^3 + \frac{87}{160}(r_L - 1)^5 + \ldots \approx \frac{60}{32}(r_L - 1)^3 \tag{12.22}$$

which equals the sum of the third-order terms in Eqs. (12.18) and (12.19).

Thus, to first order in $r_L - 1$, the net stress vanishes and the force on the electric current is balanced by the negative rate of change of momentum with time. However, as time passes and the third-order terms become important, these terms no longer balance and the net stress on the bubble becomes appreciable. At this time, 95% (57/60) of this stress is used to increase the rate of change of momentum and only 5% (3/60) pushes against the current. So, initially $\partial p_y/\partial r_L$ is negative and the momentum decreases, but after reaching the center $\partial p_y/\partial r_L$ becomes positive and the momentum increases again.

A similar expansion about $(r_L - 3)$ gives

$$\frac{p_y}{p_0} = a_0 + a_1(r_L-3) + a_2(r_L-3)^2 + a_3(r_L-3)^3 + a_4(r_L-3)^4 + a_5(r_L-3)^5 + \ldots, \tag{12.23}$$

where $a_0 = (135/64)ln3 - (29/16) \approx 0.504$, and the other coefficients have similar forms with the approximate values $a_1 = -1.608$, $a_2 = -2.337$, $a_3 = -0.473$, and $a_4 = -0.020$. With these numbers, we can reproduce the positive maximum of about 0.78 at $r_L = 2.65$. Also, by taking the derivative with respect to r_L, we can determine the rate of increase of the total momentum of the bubble:

$$\frac{1}{F_0}\frac{\partial p_y}{\partial r_L} = -1.608 - 4.674(r_L - 3) - 1.418(r_L - 3)^2 - 0.080(r_L - 3)^3 + \ldots . \tag{12.24}$$

By comparison, we can expand the exact expression for the force on the current as a finite series in powers of $(r_L - 3)$ to obtain

$$\frac{F_y}{F_0} = 3 + \left(\frac{3}{4}\right)(r_L - 3) + \left(\frac{9}{16}\right)(r_L - 3)^2 + \left(\frac{3}{32}\right)(r_L - 3)^3. \tag{12.25}$$

Consequently, during the time interval (2,3), the stress on the bubble must be the sum of these two series:

$$\frac{1}{F_0}\int T_{yy} dA_y = 1.392 - 3.924(r_L - 3) - 0.856(r_L - 3)^2 + 0.014(r_L - 3)^3 + \ldots . \tag{12.26}$$

Thus, as the trailing edge of the bubble approaches the sphere, the net stress on the entire bubble is about $1.4F_0$ with the $3F_0$ force on the current more than enough to offset the $1.6F_0$ rate of decrease of total momentum.

Next, we repeat the analysis for $r_L > 3$ when the bubble moves out from the sphere and its momentum gradually increases. Expanding the sum of the terms of Eq. (12.16) in a power series in $(r_L - 2)^{-1}$, we obtain an approximate variation for p_y outside the sphere:

$$\frac{p_y}{p_0} \approx \frac{9}{16} - \frac{3}{20}\frac{1}{(r_L - 2)^2} + \frac{19}{80}\frac{1}{(r_L - 2)^3} - \frac{3}{8}\frac{1}{(r_L - 2)^4} + \frac{3}{5}\frac{1}{(r_L - 2)^5}, \tag{12.27}$$

which shows that $p_y \to (9/16)F_0R/c = (27/64)\mathcal{E}_0/c$ as $r_L \to \infty$. Thus, not only does the bubble leave the system with energy and magnetic flux, but in each direction it takes some momentum. Although the momentum remains finite, its time rate of change approaches zero, as one can see by differentiating equation (12.27) with respect to r_L

$$\frac{1}{F_0}\frac{\partial p_y}{\partial r_L} \approx \frac{3}{10}\frac{1}{(r_L - 2)^3} - \frac{57}{80}\frac{1}{(r_L - 2)^4} + \frac{3}{2}\frac{1}{(r_L - 2)^5} - 3\frac{1}{(r_L - 2)^6}. \tag{12.28}$$

It is interesting to note that the bubble from an isolated current source carries an energy, $(3/2)\mathcal{E}_0$, and a momentum, $(3/4)^3(\mathcal{E}_0/c)$, in a single direction. This raises the question of whether there might be a conversion factor between energy and momentum given by their ratio $(32/9)c$. If so, we could use this ratio to guess the momentum carried by the linearly ramped bubble whose energy, $(6/5)(1/\Delta r)^2\ \mathcal{E}_0$, was given by Eq. (9.17). The result would be $(27/80)(1/\Delta r)^2\ \mathcal{E}_0/c$. In fact, when Eq. (12.13) is generalized by setting the lower limit of integration equal to $r_L - \Delta r - 2$ and by using the more general expression for F given by Eq. (5.9), one does obtain this value of the momentum in the y-direction when $r_L \to \infty$. Also, when $r_L \to \infty$, there is no contribution for r in the range $(r_L - \Delta r, r_L - 2)$, so that momentum is provided only by the leading and trailing sheaths of the bubble.

As shown in the right panel of Fig. 12.1, when $r_L > 3$, it is possible to surround the $(y > 0)$-portion of the bubble by a surface that does not enclose the electric current. In this case, there is no Lorentz force, and the rate of momentum increase is exactly balanced by the stress on the surrounding surface. For a closed surface consisting of hemispheres just beyond $r = r_L$ and just inside $r = r_L - 2$ that are joined by the annular end face at $y = 0$, there are three contributions to the stress: On the spherical portions of the surface, we may use Eqs. (12.5) and (12.7) to integrate the stress as follows:

$$\begin{aligned} \frac{1}{F_0} \int T_{yj} dA_j &= \pm \int_0^{\pi} \int_0^{\pi} (T_{rr} \sin\theta + T_{\theta r} \cos\theta) \sin\phi \, r^2 \sin\theta d\theta d\phi \\ &= \pm \, (r^2/4) \left[(b_r + b_\theta)^2 - 4b_\theta^2 - 3\mathcal{E}_\phi^2 \right], \end{aligned} \tag{12.29}$$

where the plus and minus signs apply to the outer and inner hemispheres, respectively. Also, $F_0 = (\pi R^2)(B_0^2/2\mu_0) = (3/4)\mathcal{E}_0/R$. There is no contribution on the spherical surface just beyond $r = r_L$ where the fields vanish, and $b_r = 0$, $b_\theta = 0$, and $\mathcal{E}_\phi = 0$. However, for the static field just inside $r = r_L - 2$, the θ-less components are $b_r = r^{-3}$, $b_\theta = (1/2)r^{-3}$, and $\mathcal{E}_\phi = 0$, so that the contribution is $-(5/16)(r_L - 2)^{-4} F_0 = -(15/64)(r_L - 2)^{-4}\mathcal{E}_0/R$, exactly.

We calculate the stress on the end face in the same way that we did when r_L was less than 2 and the bubble was headed inward. In this case, we use Eq. (12.21), but with the lower limit of the integral changed from $2 - r_L$ to $r_L - 2$. Because $r_L > 3$, we can use f_o to evaluate b_r, b_θ, and $\mathcal{E}_\phi$. Again, the integral gives a complicated algebraic expression similar to the one that we illustrated in Eq. (12.16), so we display the result as a power series in $(r_L - 2)^{-1}$. Retaining terms through sixth order, we obtain

$$\frac{1}{F_0} \int T_{yy} dA_y \approx \frac{3}{10} \frac{1}{(r_L - 2)^3} - \frac{2}{5} \frac{1}{(r_L - 2)^4} + \frac{3}{2} \frac{1}{(r_L - 2)^5} - 3 \frac{1}{(r_L - 2)^6}. \tag{12.30}$$

This series approximates the stress on the end face when $r_L \gg 2$. Adding the stress, $-(5/16)(r_L - 2)^{-4}$, from the inner hemisphere at $r = r_L - 2$, we obtain the total stress on the entire closed surface:

$$\frac{1}{F_0} \int T_{yy} dA_y \approx \frac{3}{10} \frac{1}{(r_L - 2)^3} - \frac{57}{80} \frac{1}{(r_L - 2)^4} + \frac{3}{2} \frac{1}{(r_L - 2)^5} - 3 \frac{1}{(r_L - 2)^6}. \tag{12.31}$$

As expected, this stress equals the time rate of change of momentum in Eq. (12.28). Thus, the half-bubble ultimately gains momentum from the stress on its annular end face diminished by a gradually weakening contribution of opposite polarity from its trailing surface.

Next, we consider the stress on a tight-fitting surface around the leading edge of the outgoing bubble. By letting the outer edge of this surface lie just ahead of the bubble, there can be no contribution at that surface. Likewise, for a tight-fitting surface with essentially zero thickness, there will be no contribution from the end face at $y = 0$. However, there is a contribution from the trailing portion of this closed surface, which lies just inside the bubble where $f = f_o$. Letting $r \to r_L$, we find that $b_r = 0$, $b_\theta = (3/4)r_L^{-1}$, and $\mathcal{E}_\phi = +(3/4)r_L^{-1}$. Substituting these values into Eq. (12.29) with the minus sign for this trailing surface, we obtain

$$\frac{1}{F_0}\int T_{yj}dA_j = +\frac{27}{32}. \tag{12.32}$$

Because no electric current lies within the closed surface, all of this stress must be due to the rate of increase of momentum at the leading edge of the bubble, so that $\partial p_y/\partial r_L = 27/32$. A similar analysis at the leading edge of the ingoing wave inside the sphere gives $\partial p_y/\partial r_L = -27/32$. Thus, although there can be no momentum in such infinitesimal volumes around those leading edges, nevertheless each of those edges marks a sudden change of slope where the magnitude of the momentum starts to increase. And the magnitude of this slope is $(27/32)F_0$.

Finally, we examine the stresses on a variety of surfaces behind the outgoing bubble to illustrate the relation between those stresses and the forces on the currents and rates of change of momentum that lie within those surfaces. In Fig. 12.4, these surfaces are indicated by dashed lines, numbered 1 through 6, at the time $r_L = 5$ units of R/c after the current was turned on. Surface 1 is a disk of unit radius located in the xz-plane and centered at the origin. Because the bubble has left the sphere, the disk is embedded in a uniform magnetic field, $B_0\mathbf{e_z}$, which exerts a magnetic pressure, $B_0^2/2\mu_0$, on the disk of area πR^2, corresponding to a stress $I_1 = 1$ in units of $F_0 = (B_0^2/2\mu_0)(\pi R^2)$. When surface 1 is combined with surface 6, it encloses

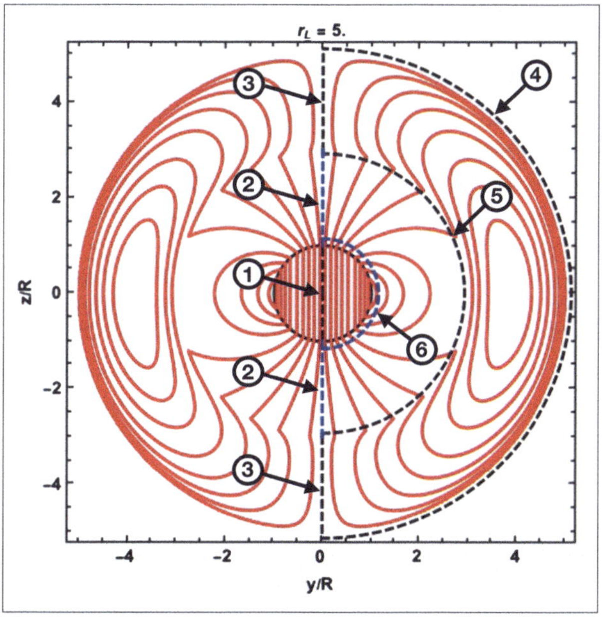

Fig. 12.4 Boundaries used for illustrating the relation between stress, force, and momentum of the field for a sudden turn-on of current on the surface of the sphere at $r = 1$

the electric current on the $(y > 0)$-side of the sphere. We have already seen in Eq. (12.12) that this current experiences a static force, $(21/16)F_0$, after the bubble has left the sphere. Consequently, $I_1 + I_6 = 21/16$, and therefore $I_6 = 5/16$.

We proceed in a similar series of steps to determine the stresses on the remaining numbered surfaces. The closed surface composed of faces 2, 5, and 6 surrounds only the static dipole field, which is a potential field with $\mathbf{\nabla x B} = \mathbf{0}$. Consequently, there is no stress on this surface and therefore $I_2 + I_5 - I_6 = 0$. (We used $-I_6$ rather than $+I_6$ because face 6 is on the left side of the volume where the y-component of the outer normal is negative.) In the discussion following Eq. (12.29), we showed that the stress on face 5 was $-(5/16)(r_L - 2)^{-4}$ when it was part of the closed surface 3-4-5, tightly fitting around the bubble. Consequently, $I_5 = (5/16)(r_L - 2)^{-4}$ and $I_2 = (5/16)\left\{1 - (r_L - 2)^{-4}\right\}$.

Surface 3 is the end face whose stress we have already calculated in Eq. (12.29). Retaining the first two terms in Eq. (12.30), we can write $I_3 \approx (3/10)(r_L - 2)^{-3} - (2/5)(r_L - 2)^{-4}$. Surface 4 lies ahead of the outgoing bubble where there are no fields or currents. Consequently, there is no stress on this surface and $I_4 = 0$. With this final I-value, we can evaluate the stress on the closed surface composed of faces 3, 4, and 5 by evaluating $I_3 + I_4 - I_5$ (again using the negative value of I_5 because face 5 is on the left side of the volume where the y-component of the outer normal is negative). The result is $(3/10)(r_L - 2)^{-3} - (57/80)(r_L - 2)^{-4}$, which matches the first two terms in the series for $\partial p_y/\partial r_L$. Consequently, $I_3 + I_4 - I_5 = \partial p_y/\partial r_L$.

Table 12.1 summarizes these results. In the right column of Table 12.1, only three of the six surface equations are independent. For example, the three concentric volumes, surrounded by surfaces 1-6, 2-5-6, and 3-4-5, correspond to the first, second, and fourth equations in the right column. The sum of the first two equations gives the third equation, which corresponds to the stress on the combined region 1-2-5. Then the sum of the third and fourth equations gives the fifth equation, which equates the stress on the entire volume to the force on the source current plus the time rate of change of the bubble's momentum. Finally, the sum of the second and fourth equations gives the sixth equation, which equates the stress on the outer two regions to the time rate of change of the bubble's momentum. The inner part of this volume contains the curl-free dipole field left behind the bubble, and therefore adds no contribution to the stress.

Table 12.1 A summary of stresses on the surfaces in Fig. 12.4

Individual faces	Closed surfaces
$I_1 = 1$	$I_1 + I_6 = 21/16$
$I_2 = (5/16) - (5/16)(r_L - 2)^{-4}$	$I_2 + I_5 - I_6 = 0$
$I_3 \approx (3/10)(r_L - 2)^{-3} - (2/5)(r_L - 2)^{-4}$	$I_1 + I_2 + I_5 = 21/16$
$I_4 = 0$	$I_3 + I_4 - I_5 = \partial p_y/\partial r_L$
$I_5 = (5/16)(r_L - 2)^{-4}$	$I_1 + I_2 + I_3 + I_4 = 21/16 + \partial p_y/\partial r_L$
$I_6 = 5/16$	$I_2 + I_3 + I_4 - I_6 = \partial p_y/\partial r_L$

$\partial p_y/\partial r_L \approx (3/10)(r_L - 2)^{-3} - (57/80)(r_L - 2)^{-4}$

From these examples, it is clear that there can be a net stress on a closed surface only if it contains the source current or the bubble. However, the inverse is not true. By comparing Eqs. (12.18) and (12.19), we found that immediately after the current was turned on, the force on the current equalled the negative rate of increase of momentum, so that the net stress vanished for a while. Also, the leading and trailing edges of the bubble are places where the time rate of change of momentum changes suddenly. In the next chapter, we shall apply these ideas to the fields of two spherical current sources.

Chapter 13
The Interaction of Two Sources

Next, we consider what happens when two spherical currents are turned on simultaneously according to a clock located midway between them. We suppose that the spheres have identical radii, R, and current densities, $\mathbf{J}$ (given essentially by Eq. (2.1)), and that they are separated by the distance, $2a$ (where $a > R$). The simultaneity can be achieved by sending a signal from the midpoint of the line joining the spheres. When observers at the centers of the two spheres see this signal, they can set their clocks and send out their own signals to turn on the currents in their respective spheres. Thus, after an additional delay of R/c, each current will suddenly turn on, sending waves both inward and outward from each sphere. In this way, each source would generate its own toroidal bubble, dipole field, and uniformly magnetized sphere.

If the two sources are turned on suddenly at time $t = R/c$, the resulting bubbles will reach the mid-plane at the later time $t = a/c$, or $r_L = a$ in the convention of the previous chapters (time in units of R/c and distance in units of R). They subsequently pass through each other, leaving the composite field of two dipoles in their common wake. As the bubbles expand further, the region of composite field also expands, eventually encompassing the spheres themselves as the bubbles pass the opposite current sources and escape from the source region.

13.1 Parallel Dipoles

Figure 13.1 illustrates the field-line topology in the plane of the dipole axes when the dipole moments are parallel to each other and perpendicular to the line joining the two spheres. We have chosen $a = 8$ and computed field lines for $r_L = 12$ and

Electronic Supplementary Material The online version of this chapter (https://doi.org/10.1007/978-3-030-40264-8_13) contains supplementary material, which is available to authorized users.

N. R. Sheeley, Jr., *Transient Magnetic Fields*,
https://doi.org/10.1007/978-3-030-40264-8_13

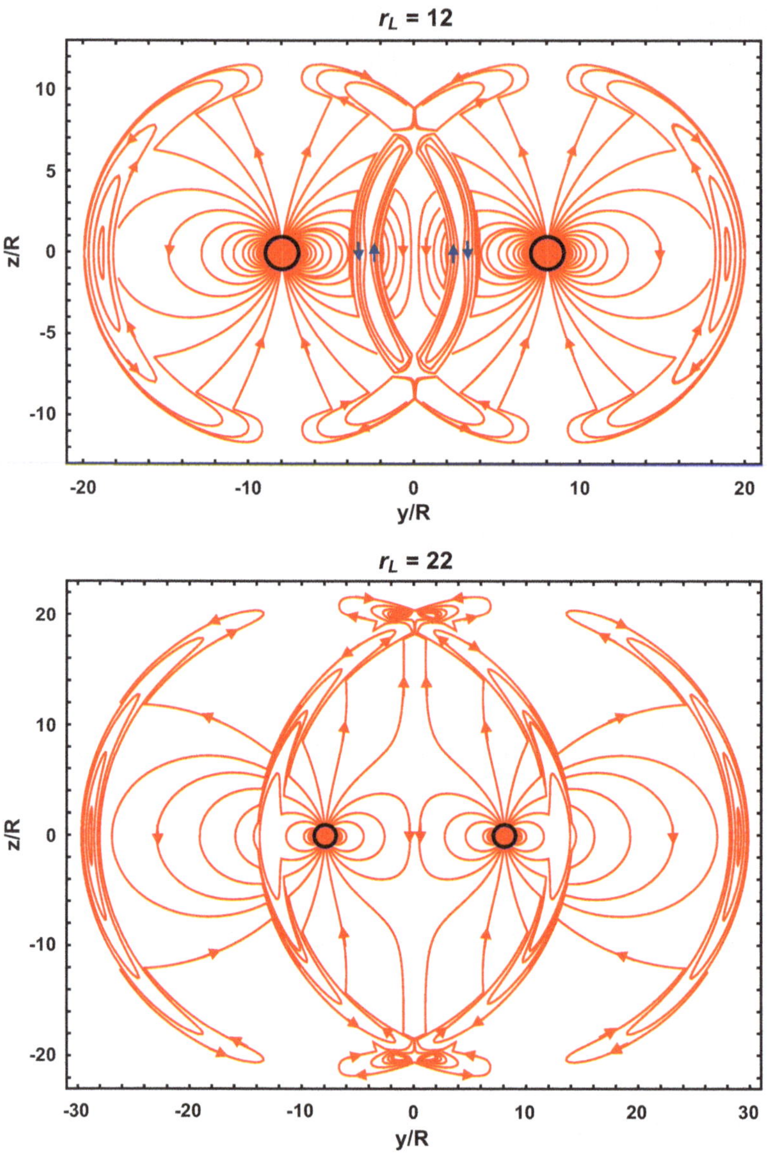

Fig. 13.1 Field lines for two identical current sources with parallel dipole moments, showing flattened composite fields at times $r_L = 12$ (top) and $r_L = 22$ (bottom). The distance between the spheres is $2a = 16$

$r_L = 22$, corresponding to times before and after the bubbles have passed their opposite sources, respectively.

For this parallel configuration, there are no magnetic field components normal to the mid-plane and no electric field components in the plane. In effect, the field on each side of the mid-plane evolves as if the plane were a superconducting wall and the other source were not present. Consequently, we can use this two-source configuration to calculate the force transmitted to a superconducting wall when the bubble from a single source reflects off it.

We adopt a rectangular coordinate system in which the wall is represented by the plane at $y = 0$ and the sphere is located to its right on the y-axis where $y = a$. The net force, F_y, on a large, but very thin, box enclosing the wall is given by

$$F_y = \int_{box} T_{yj} dA_j - \frac{\partial}{\partial t}\left[\frac{1}{c^2}\int_{box} S_y dV\right], \tag{13.1}$$

where T_{yj} is the Maxwell stress tensor given by

$$T_{yj} = \frac{B_y B_j}{\mu_0} - \delta_{yj}\left(\frac{B^2}{2\mu_0}\right) + \epsilon_0 E_y E_j - \delta_{yj}\left(\frac{1}{2}\epsilon_0 E^2\right), \tag{13.2}$$

and $S_y = (\mathbf{E}\times\mathbf{B}/\mu_0)\cdot\mathbf{e_y}$ is the Poynting energy flux in the y-direction. In the limit of a very thin box, the volume integral vanishes and the surface integral has non-vanishing contributions only on the two faces for which $j = y$. Because the wall is superconducting, there are no fields on the face where $y < 0$, and only the face where $y > 0$ contributes. In effect, F_y is the force on the induced currents on the surface of the wall. Taking this face arbitrarily close to the wall, we may express the force, F_y, as

$$F_y = \int_{\text{wall}} T_{yy} dA_y = -\int_{\text{wall}}\left(\frac{B_x^2 + B_z^2}{2\mu_0}\right) dA_y + \int_{\text{wall}}\left(\frac{1}{2}\epsilon_0 E_y^2\right) dA_y, \tag{13.3}$$

where the field components, B_x, B_z, and E_y, are evaluated at the wall ($y = 0$) and dA_y is positive in the $+y$-direction. (If we were calculating the force on the sphere at $y = a$ instead of the wall at $y = 0$, we would have chosen $dA_y < 0$ to make the outer normal in the $-y$-direction.) Because all of the field components are squared, Eq. (13.3) indicates that the "flattened" magnetic field pushes inward against the wall, whereas the electric field of the bubble pulls the wall outward in the direction of positive y.

To evaluate F_y, we work in the spherical coordinate system centered at one of the spheres, say the sphere at $y = +a$, where B_r, B_θ, and E_ϕ are known, and recognize that the other sphere simply doubles the contributions of B_x, B_z, and E_y in Eq. (13.3). Because these components are squared, this accounts for a factor of 4 in the resulting expression for F_y. Projecting B_r, B_θ, and E_ϕ onto the xyz coordinate axes, we have $B_x = B_r \sin\theta \cos\phi + B_\theta \cos\theta \cos\phi$, $B_z = B_r \cos\theta - B_\theta \sin\theta$, and

$E_y = -E_\phi \cos\phi$. Next, we introduce cylindrical coordinates (ρ, α) centered around the origin in the xz-plane and recognize that the areal element, $dA_y = \rho d\rho d\alpha$. Now, $\rho d\rho = r dr$ because $r^2 = \rho^2 + a^2$. Also, α is related to θ and ϕ by the equations $\cos^2\theta = p \cos^2\alpha$ and $\sin^2\theta \cos^2\phi = p \sin^2\alpha$ where $p = 1 - (a/r)^2$. Integrating over α in the range $(0, 2\pi)$ produces factors of π, which combine with our unit of length squared, R^2, and the magnetic pressure, $B_0^2/2\mu_0$, to give the unit of force $F_0 = (B_0^2/2\mu_0)(\pi R^2)$. Then, after a little algebra, we obtain

$$F_y/F_0 = -4\int_a^{r_L} \left[\{p\, b_r - (1-p)\, b_\theta\}^2 + b_\theta^2 - p\, \mathcal{E}_\phi^2 \right] r dr, \tag{13.4}$$

where $p(r, a) = 1 - (a/r)^2$ and $F_0 = (\pi R^2)(B_0^2/2\mu_0)$. The quantities b_r, b_θ, and $\mathcal{E}_\phi$ are the θ-less field components given by Eqs. (12.7) and (12.8), and $f(r, r_L)$ is the normalized vector potential given by Eq. (2.10). Because no force is exerted until the bubble arrives at the wall, the lower and upper limits of the integral are $r = a$ and $r = r_L$, respectively. When r_L lies in the range $(a, a+2)$, only the transient part of the disturbance is in contact with the wall, so that $f(r, r_L)$ is given by Eq. (2.10) with the minus sign (i.e., $f = f_o$) and the limits become a and r_L. Later, when $r_L > a + 2$, the force has contributions from both the composite dipole and the bubble. In this case, $f = 1$ for r in the range $(a, r_L - 2)$, and $f = f_o$ for r in the range $(r_L - 2, r_L)$.

Thus, the force exerted by the composite dipole is

$$F_y/F_0 = \begin{cases} 0, & r_L < a+2, \\ -4\int_a^{r_L-2} \left[\{p\, b_r - (1-p)\, b_\theta\}^2 + b_\theta^2 - p\, \mathcal{E}_\phi^2 \right] r dr, & r_L > a+2, \end{cases} \tag{13.5}$$

where b_r, b_θ, and $\mathcal{E}_\phi$ are obtained by setting $f = 1$ in Eqs (12.8a–c). In this case, $b_r = r^{-3}$, $b_\theta = (1/2)r^{-3}$, and $\mathcal{E}_\phi = 0$, which leads to the exact polynomial expression

$$\frac{F_y}{F_0} = -\left(\frac{3}{8a^4}\right)\left[1 - \frac{10}{3}\left(\frac{a}{r_L - 2}\right)^4 + \frac{16}{3}\left(\frac{a}{r_L - 2}\right)^6 - 3\left(\frac{a}{r_L - 2}\right)^8\right], \tag{13.6}$$

when $r_L > a + 2$. Thus, the force exerted by the dipole field has a transient component even though the dipole lacks an electric field. This time dependence is a consequence of the expanding area of the wall that is subject to the dipole magnetic pressure, as evidenced by the upper limit, $r_L - 2$, in the integral for F_y/F_0 in Eq. (13.5).

Eventually, as r_L becomes much greater than $a + 2$, F_y/F_0 approaches the static value $-(3/8)a^{-4}$, whose magnitude is also the force with which the stressed dipole field pushes back against the sphere. If the sphere were not held in place, it would move in the positive y-direction and acquire an amount of kinetic energy equal to

$\int_a^\infty F_y da = (3/8) \int_a^\infty a^{-4} da = (1/8)a^{-3}\ F_0 R = (3/32)a^{-3}\ \mathcal{E}_0$, with a expressed in units of R. The other sphere would feel an equal force of repulsion and acquire the same kinetic energy as it recedes in the $-y$-direction. Although this suggests that the flattened field might have an extra energy of $(3/16)a^{-3}\ \mathcal{E}_0$, we shall see in the next chapter that this is only part of the energy. When all of the energy is considered, the energy stored in the flattened field is actually less than the energy in the field of the anti-parallel dipoles.

In contrast, the force exerted by the bubble is

$$F_y/F_0 = \begin{cases} 0, & r_L < a, \\ -4 \int_a^{r_L} \left[\{p\, b_r - (1-p)\, b_\theta\}^2 + b_\theta^2 - p\, \mathcal{E}_\phi^2 \right] r dr, & a < r_L < a+2, \\ -4 \int_{r_L-2}^{r_L} \left[\{p\, b_r - (1-p)\, b_\theta\}^2 + b_\theta^2 - p\, \mathcal{E}_\phi^2 \right] r dr, & r_L > a+2, \end{cases} \tag{13.7}$$

where b_r, b_θ, and $\mathcal{E}_\phi$ are obtained by setting $f = f_o$ in Eqs. (12.8a–c). This substitution leads to more difficult integrals. Although these integrals can be evaluated exactly, they give very complicated expressions with logarithmic terms and long series of algebraic terms. Rather than reproducing these expressions here, we will plot their values as a function of time, r_L, and look for relations in the resulting graphs.

Figure 13.2 shows the results plotted for $a = 1$, 1.5, 2, and 4 without the minus sign that indicates a force against the wall in the $-y$-direction. So we are really plotting the force obtained by selecting $dA_y < 0$ in Eq. (13.3). As expected, the force vanishes until the leading edge of the bubble arrives at the wall when $r_L = a$. Then, as r_L moves through the range $(a, a+2)$, the force increases, decreases, and then increases again as the leading, central, and trailing parts of the bubble reach the wall. As r_L exceeds $a+2$, the contribution of the bubble falls toward zero while the contribution of the composite dipole field increases toward its asymptotic value. Progressing from the upper left panel to the lower right panel, we see that the asymptotic contribution of the composite dipole decreases rapidly with a and is essentially gone when $a = 4$. Thus, when the sphere is located close to the wall, the initially strong force of the bubble is quickly replaced by the static force of the composite dipole field. However, if the sphere is located farther than about 2 radii from the wall, the static field never gets very strong and the only significant contribution comes from the bubble.

In Fig. 13.2, the consecutive ups and downs that occur while the bubble is passing through the interval $(a, a+2)$ resemble the wavy shape of a cubic polynomial. This suggests that we may be able to find a relatively simple cubic polynomial that approximates the force of the bubble during this time interval. For this purpose, we begin by defining $x = r_L - (a+1)$ and consider small variations about the midpoint of the interval $(a, a+2)$:

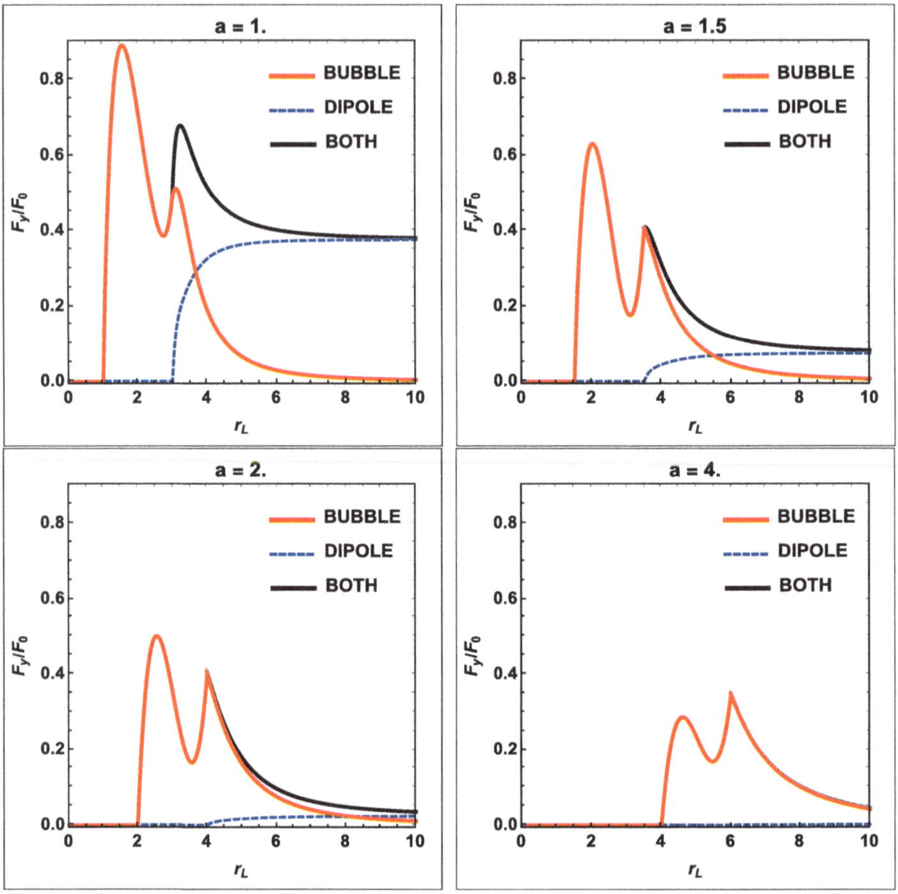

Fig. 13.2 The force that a bubble and its trailing dipole field exert against a superconducting wall as a function of elapsed time, r_L. The curves are plotted for four distances, a, between the center of the sphere and the wall: upper left ($a = 1.0$ with the sphere just tangent to the wall), upper right ($a = 1.5$), lower left ($a = 2.0$), and lower right ($a = 4.0$). The force, F_y, is expressed in units of the force $F_0 = (\pi R^2)(B_0^2/2\mu_0)$

$$-F_y/F_0 \equiv G(x, a) = c_0(a) - c_1(a)x + c_2(a)x^2 + c_3(a)x^3, \tag{13.8}$$

where the coefficients c_0 through c_3 are power series in the quantity $1/a$, and the symbol, $G(x, a)$, represents the force of the bubble against the wall, expressed as a function of x and a. The idea here is to locate the peaks of this cubic by setting $\partial G(x, a)/\partial x = 0$ and solving the resulting quadratic equation for x. Ideally, the peaks would be equidistant from $x = 0$ where $\partial^2 G(x, a)/\partial x^2 = 0$. This would require $c_2 = 0$, in which case the peaks would be offset from the center by the distance $(c_1/3c_3)^{1/2}$. (To avoid the appearance of a minus sign under this radical, we introduced $-c_1$ in Eq. (13.8), rather than $+c_1$.)

When we expanded the exact expression for F_y/F_0 in powers of x, we found that c_2 was not zero, which meant that the center was offset slightly from $r_L = a + 1$. Consequently, we changed the point of expansion from $a+1$ to $a+1+\frac{1}{2}a^{-1}-\frac{19}{6}a^{-2}$, whose extra terms were selected to make $c_2(a)$ vanish through terms of order a^{-3}. With this change, the coefficients become

$$c_0(a) = \frac{3}{2a}\left(1 - \frac{9}{4}a^{-1} + \frac{21}{10}a^{-2} + \frac{143}{12}a^{-3}\right), \tag{13.9a}$$

$$c_1(a) = \frac{3}{2a}\left(4a^{-1} - \frac{21}{2}a^{-2} + \frac{71}{5}a^{-3}\right), \tag{13.9b}$$

$$c_2(a) = -\frac{3}{2a}\left(\frac{249}{8}a^{-3}\right), \tag{13.9c}$$

$$c_3(a) = \frac{3}{2a}\left(1 + \frac{3}{2}a^{-2} - \frac{52}{3}a^{-3}\right), \tag{13.9d}$$

with $x = r_L - (a + 1 + \frac{1}{2}a^{-1} - \frac{19}{6}a^{-2})$. Together, Eqs. (13.8) and (13.9) provide a good fit to F_y/F_0 for values of $a \geq 6$, as shown in Fig. 13.3.

Now, for $a \gg 1$, the central height of $G(x, a)$ is $G(0, a) = c_0(a) \approx (3/2a)\{1 - (9/4)a^{-1}\}$. The locations of the adjacent maximum and minimum values of $G(x, a)$ are obtained by setting $\partial G/\partial x = 0$ with the result that $x \approx \pm(c_1/3c_3)^{1/2} \approx \pm(4/3a)^{1/2}$ units of R on each side of center. The corresponding heights extend approximately $G(x, a) - c_0(a) \approx -c_1 x + c_3 x^3 \approx \pm 72(3a)^{-5/2}$ units of F_0 above and below the central height. Thus, as the distance between the spheres increases,

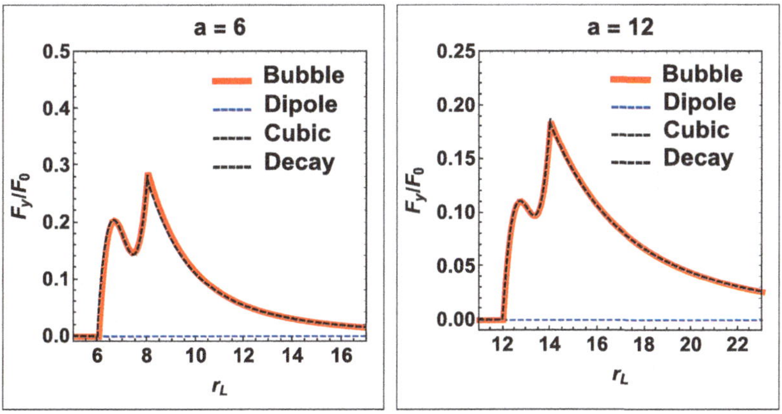

Fig. 13.3 The force that a bubble and its trailing dipole field exert against a superconducting wall as a function of elapsed time, r_L, for $a = 6$ (left) and $a = 12$ (right), showing the near coincidence between the exact (red) and approximate (black dashed) expressions when $a \geq 6$. The force, F_y, is expressed in units of $F_0 = (\pi R^2)(B_0^2/2\mu_0)$, and is plotted without the minus sign

the center of the profile moves closer to $r_L = a + 1$, and its height decreases. Also, the adjacent peak and dip move closer together and shrink in size, eventually coalescing at $r_L = a + 1$ to form an inflection point in the temporal profile, as given by the lowest-order approximation

$$-F_y/F_0 \approx \frac{3}{2a}\left[1 + (r_L - a - 1)^3\right]. \tag{13.10}$$

In this case, the profile varies from 0 at $r_L = a$, to $3/2a$ at the midpoint where $r_L = a + 1$, and then to $3/a$ at the right end of the interval where $r_L = a + 2$.

A similar approach can be applied to the subsequent decay of the bubble force after the time $r_L = a + 2$. Expressing the second integral in Eq. (13.7) as a power series in a/r_L, we obtain

$$G(r_L, a) \approx \left(\frac{3}{2a}\right)\left(\frac{a}{r_L}\right)^3 \left\{k_0 + k_1\left(\frac{a}{r_L}\right) + k_2\left(\frac{a}{r_L}\right)^2 + k_3\left(\frac{a}{r_L}\right)^3 + k_4\left(\frac{a}{r_L}\right)^4\right\}, \tag{13.11}$$

where the coefficients of this expansion are

$$k_0(a) = 1 + \frac{4}{5}a^{-2}, \tag{13.12a}$$

$$k_1(a) = 3a^{-1}, \tag{13.12b}$$

$$k_2(a) = 1 + \frac{42}{5}a^{-2}, \tag{13.12c}$$

$$k_3(a) = 5a^{-1}, \tag{13.12d}$$

$$k_4(a) = 18a^{-2}, \tag{13.12e}$$

accurate through terms of $\mathcal{O}(a^{-2})$. (We retained higher-order terms for these coefficients, but found that they did not affect the result for $a \geq 6$, provided that all five terms in the expansion of $G(r_L, a)$ were retained in Eq. (13.11).)

The result is plotted in Fig. 13.3 as the segment of the black dashed curve with $r_L \geq a + 2$. Like the cubic segment with r_L in the range $(a, a + 2)$, this decaying segment is virtually coincident with the corresponding segment of the exact force plotted in red. When $G(r_L, a)$ is expanded in powers of $a/(r_L - 2)$, the resulting coefficients are very nearly the same as the ones above, except that k_1 and k_3 are negative. However, the approximation in terms of $a/(r_L-2)$ is less accurate than the approximation in terms of a/r_L, even when the higher-order terms in $k_0(a)$ through $k_4(a)$ are included (i.e., the $a/(r_L - 2)$ approximation did not begin to show a close agreement with the exact expression until $a{\sim}12$).

Up to this point, we have regarded the field of the two parallel dipoles as equivalent to the field of a single dipole in front of a superconducting wall, and we have interpreted the resulting stress in terms of the force on the induced currents on the surface of the wall. However, as an alternative approach, we can regard the xz-plane as the left boundary of a closed surface that encloses the sphere at $y = a$ and all of the fields in the region where $y > 0$. Then we can use the stress on this surface to determine the force on the sphere and the rate of momentum increase of the bubbles passing through this region. For this purpose, we simply reverse the polarities of the stresses that we have already calculated and study their effects on the sphere at $y = a$ and the bubbles passing through the $(y > 0)$-region as a function of time.

For $r_L < a$, the bubbles have not reached the mid-plane yet, and the net force and momentum all vanish. For r_L in the range $(a, a + 2)$, the bubbles have crossed the mid-plane, and the stress is given by the second line of Eq. (13.7) (with the leading minus sign changed to a plus sign). This term corresponds to the initial, cubic-shaped profile in Figs. 13.2 and 13.3, and to the approximation given in Eqs. (13.8) and (13.9). At this time, the right-going bubble has not reached the sphere at $y = a$, so none of the stress exerts a force on the current on that sphere; all of the stress goes into increasing the momentum of the bubbles in the $(y > 0)$-region.

What is the distribution of momentum in this region? To answer this question, we refer to the fields plotted in Fig. 13.1. Prior to $r_L = a$, the bubbles would not have reached the mid-plane, and the rightward component of momentum, p_y, from the sphere at $y = a$ would cancel the leftward component, giving a net momentum of 0. However, after $r_L = a$, part of the leftward bubble has crossed the mid-plane, reducing the amount of leftward momentum in the $(y > 0)$-region by this amount, p_c. Also, an equal amount of momentum from the sphere at $y = -a$ has crossed into the $(y > 0)$-region, giving a net change of $2p_c$. As a result, the stress equation becomes

$$\int T_{yy} dA_y = \dot{p}_y - (\dot{p}_y - \dot{p}_c) + \dot{p}_c = 2\dot{p}_c, \tag{13.13}$$

where $\dot{p}_c = \partial p_c / \partial r_L$ and $\dot{p}_y = \partial p_y / \partial r_L$. In our previous interpretation, we would have called p_c the momentum carried by the bubble as it was reflected from the wall.

Returning to the situation for r_L in the range $(a, a + 2)$, we now understand that the stress on the mid-plane at this time is just $2\dot{p}_c$, which we set equal to the function $G(x, a)$, given by Eq. (13.8) with $x \approx r_L - (a + 1)$:

$$2\frac{\partial p_c}{\partial r_L} = c_0(a) - c_1(a)(r_L - a - 1) + c_2(a)(r_L - a - 1)^2 + c_3(a)(r_L - a - 1)^3, \tag{13.14}$$

where the coefficients, c_0 through c_3 are given approximately by Eq. (13.9). Solving this differential equation for $p_c(r_L, a)$, and neglecting terms of order a^{-2} and greater, we obtain

$$2p_c \approx k + \frac{3}{2a}\left[(r_L - a - 1) + \frac{1}{4}(r_L - a - 1)^4\right], \tag{13.15}$$

where k is a constant of integration. We can estimate the value of k by requiring that $p_c = 0$ when $r_L = a$. Consequently, $k \approx 9/8a$, and Eq. (13.15) reduces to

$$p_c(r_L, a) \approx \frac{9}{16a}\left[1 + \frac{4}{3}(r_L - a - 1) + \frac{1}{3}(r_L - a - 1)^4\right]. \tag{13.16}$$

From this equation, we see that $p_c(a, a) = 0$ (as required by the choice of k), that $p_c(a + 1, a) = 9/16a$, and that $p_c(a + 2, a) = 3/2a$. In addition, by taking the derivative of Eq. (13.16) with respect to r_L and doubling the result, we recover the low-order cubic relation for the stress given in Eq. (13.10). For a solution that is valid close to the starting point, $r_L = a$, we could have expanded Eq. (13.8) in powers of $r_L - a$ and performed a similar analysis, constraining the constant of integration by $p_c(a, a) = 0$ and obtaining

$$p_c(r_L, a) \approx \frac{9}{8a}(r_L - a)^2\left[1 - \frac{2}{3}(r_L - a) + \frac{1}{6}(r_L - a)^2\right]. \tag{13.17}$$

But a little algebra shows that this equation is identical to Eq. (13.16).

For $r_L > a + 2$, the composite dipole field and the bubble both exert stress on the mid-plane. The dipole contribution, G_{dip}, is given in Eq. (13.6) without the minus sign and the contribution of the bubble, G_{bub}, is given in the third line of Eq. (13.7), again without the minus sign. If $r_L < 2a - 1$, the rightward-traveling bubble from the sphere at $y = -a$ has not yet reached the sphere at $y = a$, so that the total stress on the mid-plane will be used to increase the momentum of that bubble. In this case, $2\dot{p}_c = G_{dip} + G_{bub}$. However, after $r_L = 2a + 3$, the bubble will have passed the sphere at $y = a$, and the static dipole field left behind will exert a net force, $F_y = 3/8a^4 F_0$, on the current on the surface of that sphere. This force and the momentum rate, $2\dot{p}_c$, will share the total stress on the mid-plane, and the stress equation will become $3/8a^4 + 2\dot{p}_c = G_{dip} + G_{bub}$. The extra stress balances the leading term in the series for G_{dip} (Eq. (13.6) without the leading minus sign), so that $2\dot{p}_c \approx G_{bub}$. (However, recall from Figs. 13.2 and 13.3 that G_{dip} is very small for $a > 2$, so that G_{dip} can be neglected even before the bubble arrives at the sphere, unless $a \sim 1$.)

We use an approximate method to solve the equation $2\dot{p}_c = G_{bub}$, obtaining the stress, G_{bub}, from Eq. (13.11) whose coefficients are given by Eq. (13.12). Neglecting terms of order a^{-2} and greater, we find

$$2\dot{p}_c = \frac{3}{2a}\left(\frac{a}{r_L}\right)^3. \tag{13.18}$$

Then, solving for p_c, we obtain

$$2p_c = k + \frac{3}{4}\left(\frac{a}{r_L}\right)^2, \tag{13.19}$$

where k is the constant of integration. We evaluate this constant by letting $r_L \to \infty$ and requiring the asymptotic value of p_c to be 9/16, as given in Eq. (12.27) for a single source. The result is

$$p_c = \frac{9}{16}\left[1 - \frac{2}{3}\left(\frac{a}{r_L}\right)^2\right]. \tag{13.20}$$

If we return to the series expansion for G_{bub} and include terms through order $(a/r_L)^5$, but still neglect terms of order a^{-2} in the expansion coefficients, we obtain

$$p_c = \frac{9}{16}\left[1 - \frac{2}{3}\left(\frac{a}{r_L}\right)^2 - \frac{4}{3a}\left(\frac{a}{r_L}\right)^3 - \frac{1}{3}\left(\frac{a}{r_L}\right)^4 - \frac{4}{3a}\left(\frac{a}{r_L}\right)^5\right], \tag{13.21}$$

which simplifies to

$$p_c = \frac{9}{16}\left[1 - \frac{2}{3}\left(\frac{a}{r_L}\right)^2 - \frac{1}{3}\left(\frac{a}{r_L}\right)^4\right] \tag{13.22}$$

when the factors $4/3a$ are neglected. Even when the residual terms in G_{dip} were included, none of them survived the final neglect of terms of order a^{-2} and higher in the expansion coefficients. Using the wall terminology, this $+y$-directed momentum is due solely to the stress exerted when the bubble reflected off the wall, and does not include the usually small push of the composite dipole field.

At this point, it is useful to combine the approximate expressions given by Eqs. (13.16) and (13.22) into a single equation for the momentum, p_c, as a function of time, r_L. The result is

$$\frac{p_c}{p_0} = \begin{cases} 0, & 0 < r_L < a, \\ (9/16a)\left[1 + \frac{4}{3}(r_L - a - 1) + \frac{1}{3}(r_L - a - 1)^4\right], & a < r_L < a+2, \\ (9/16)\left[1 - \frac{2}{3}\{a/(r_L - 1)\}^2 - \frac{1}{3}\{a/(r_L - 1)\}^4\right], & r_L > a+2. \end{cases} \tag{13.23}$$

We replaced r_L by $r_L - 1$ in Eq. (13.22) before inserting it into the third part of Eq. (13.23). This is justified when $a \gg 1$ because $r_L > a+2$ and therefore $r_L \gg 1$. This change has the advantage of making p_c continuous across the boundary at $r_L = a + 2$, at least to first order in $1/a$, and it also retains the asymptotic value of $p_c = 9/16$ as $r_L \to \infty$.

Figure 13.4 plots the result for two values of a, showing that the match at $r_L = a + 2$ improves as a increases. In this figure, the changing slopes during the time interval $(a, a+2)$ are hardly noticeable, and the evolution is dominated by the gradual approach to the asymptotic value, $p_c = 9/16 \approx 0.56$.

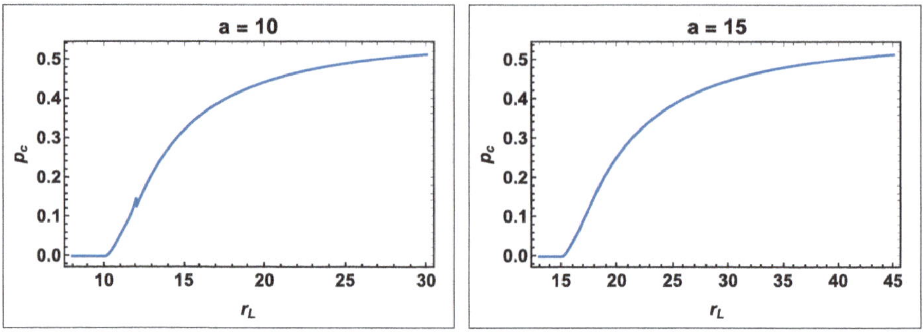

Fig. 13.4 The momentum, p_c, computed using the approximations in Eq. (13.23), and plotted versus time, r_L. These plots show the asymptotic approach to $p_c = 9/16 \approx 0.56$, and the improved continuity at $r_L = a + 2$ as the half-separation, a, between the two spheres increases from 10 to 15

13.2 Anti-parallel Dipoles

If the current source were oriented so that its dipole moment made an oblique angle with respect to the wall, then presumably the resulting fields could be expressed as a sum of the fields from the oblique source and its oblique image behind the wall. Next, we consider the special case for which the dipole moments of the two sources are anti-parallel and perpendicular to the line joining the spheres. Figure 13.5 illustrates this case.

In the xz plane between these oppositely directed sources, the electric field lies in the plane and the magnetic field is perpendicular to it. As the oppositely directed meridional magnetic fields of the colliding bubbles overlap, their field lines circulate around an O-type neutral point. However, these circulating magnetic field lines move with the bubbles, and leave an X-type neutral point of reconnected field lines in their wake. As the bubbles sweep past the opposite spheres, the O-type neutral points split in a process that is reminiscent of the motion of a plane wave past a dipole field that we examined in Sect. 8.2 and Fig. 8.4.

When $r_L > a + 2$, the X-type neutral point separates the field into "reconnected" field lines that join the two spheres, and "non-reconnected" field lines that loop back to the individual spheres. We can calculate the amount of flux that is linked between the two spheres by evaluating the integral $\int_a^{r_L - 2} B_y dA_y$ when $r_L > a + 2$. The procedure is similar to the one that we used to derive the force in Eq. (13.4), except that the force integral extended over 2π radians of azimuth in that case. Here, the flux integral spans only π radians, corresponding to the upper—or lower-half of the xz plane. For the upper half of the plane, we obtain the flux, $\Phi = -2B_0 \int_a^{r_L-2} \int_0^{\pi} (b_r + b_\theta) \sin\theta \cos\theta \sin\phi \; r dr d\alpha$. Substituting the relations $\sin\theta \sin\phi = a/r$ and $\cos\theta = [1 - (a/r)^2]^{1/2} \sin\alpha$ into this equation and performing the integration over α and r, we finally obtain

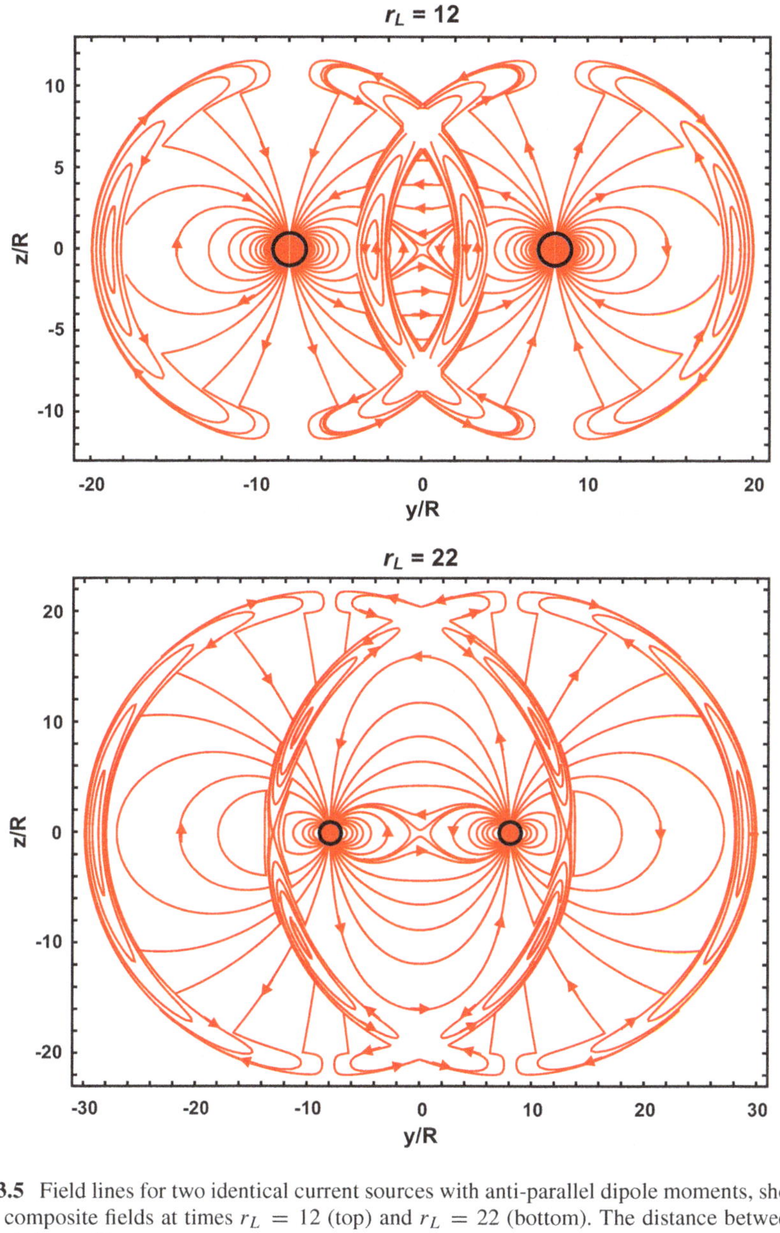

Fig. 13.5 Field lines for two identical current sources with anti-parallel dipole moments, showing linked composite fields at times $r_L = 12$ (top) and $r_L = 22$ (bottom). The distance between the spheres is $2a = 16$

$$\frac{\Phi}{\Phi_0} = \frac{2}{\pi a}\left[1 - \left(\frac{a}{r_L - 2}\right)^2\right]^{3/2}, \tag{13.24}$$

where $\Phi_0 = \pi R^2 B_0$. Like Eq. (13.6), Eq. (13.24) depends on the quantity, $(r_L - 2)/a$, whose time dependence corresponds to the increasing size of the region crossed by field lines that join the two spheres. Thus, the amount of interconnected flux is zero when $r_L = a + 2$, and increases to $(2/\pi a)\Phi_0$ as $r_L/a \to \infty$, while the remaining flux that loops back to each sphere approaches $[1 - (2/\pi a)]\Phi_0$.

If we evaluate the stress on a closed surface consisting of the xz-plane and a large region enclosing the sphere at $y = a$ and located in the field-free region beyond the advancing waves, we would obtain

$$\int T_{yy} dA_y = -\int_{\text{plane}} \left[\frac{B_y^2}{2\mu_0} - \frac{1}{2}\epsilon_0 E_x^2 \right] dA_y$$

$$= -4F_0 \int_a^{r_L} (1-p) \left[p(b_r + b_\theta)^2 - 2\mathcal{E}_\phi^2 \right] r dr, \tag{13.25}$$

where $p = 1 - (a/r)^2$, and b_r, b_θ, and $\mathcal{E}_\phi$ are the θ-less field components given in Eq. (12.8). This equation gives the contribution of the bubble when $r_L < a + 2$. However, when $r_L > a + 2$, the lower limit must be changed to $r_L - 2$ so that the integration spans the usual range, $(r_L - 2, r_L)$, for the bubble.

Also, when $r_L > a + 2$, the dipole contribution is obtained by setting $b_r = r^{-3}$, $b_\theta = (1/2)r^{-3}$, and $\mathcal{E}_\phi = 0$ and changing the upper limit to $r_L - 2$. In this case, we obtain the exact expression

$$F_y/F_0 = -\frac{3}{8a^4} \left[1 - 4\left(\frac{a}{r_L - 2}\right)^6 + 3\left(\frac{a}{r_L - 2}\right)^8 \right], \tag{13.26}$$

whose leading minus sign indicates a force in the $-y$-direction. Consequently, after the transient, the sphere at $y = a$ is being pulled in the $-y$-direction with a magnitude $(3/8)a^{-4}F_0$. Of course, the sphere at $y = -a$ is pulled in the $+y$-direction by a force of the same magnitude. Thus, after the transients, there is an attractive force between anti-parallel dipoles and a repulsive force of the same magnitude between the parallel dipoles. However, this magnitude is $(3/8)a^{-4}$, which is $0.023F_0$ for $a = 2$, and decreases rapidly as a becomes larger. Thus, the static force between the two spheres is almost negligible unless $a \sim 1$ and the spheres are virtually in contact.

The transient force has a brief rise and fall as the bubble crosses the mid-plane during the time interval $(a, a + 2)$. However, this force ends up being mainly repulsive for both the parallel and anti-parallel dipoles. In fact, when $r_L = a + 2$ the force on the sphere located at $y = a$ can be expressed as a power series in $1/a$. The result is

$$\frac{F_y}{F_0} = +\frac{3}{a}\left(1 - \frac{4}{a} + \frac{68}{5}\frac{1}{a^2} - \frac{221}{5}\frac{1}{a^3} + \frac{948}{7}\frac{1}{a^4} - \frac{394}{a^5} + \dots\right) \tag{13.27}$$

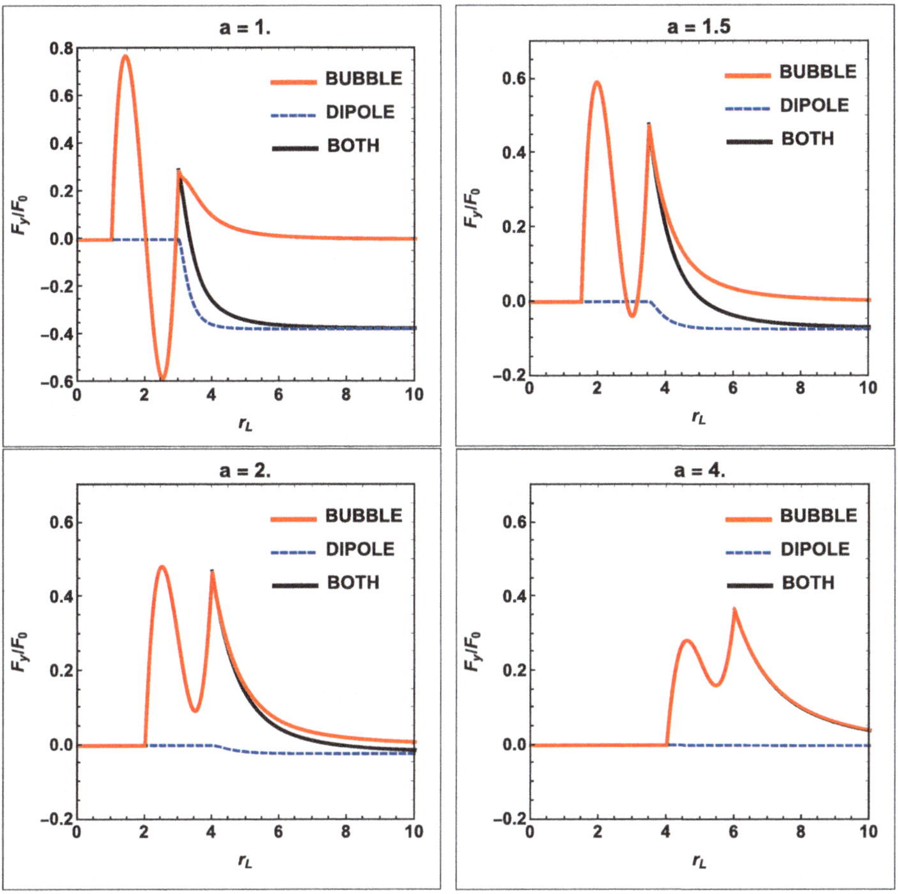

Fig. 13.6 The stress that a bubble and its trailing dipole field exert on the mid-plane at $y = 0$, as a function of elapsed time, r_L, for anti-parallel dipole moments. As in Fig. 13.2, the curves are plotted for four half-separation distances, a, between the spheres: upper left ($a = 1.0$), upper right ($a = 1.5$), lower left ($a = 2.0$), and lower right ($a = 4.0$). The force, F_y, is expressed in units of the force $F_0 = (\pi R^2)(B_0^2/2\mu_0)$

for parallel dipoles, and

$$\frac{F_y}{F_0} = +\frac{3}{a}\left(1 - \frac{4}{a} + \frac{72}{5}\frac{1}{a^2} - 47\frac{1}{a^3} + \frac{1000}{7}\frac{1}{a^4} - \frac{414}{a^5} + \dots\right) \tag{13.28}$$

for anti-parallel dipoles. For $a \gg 1$, these expressions are nearly the same.

We can see this behavior by comparing the plots for anti-parallel dipole fields in Fig. 13.6 with the previous plots for the parallel dipoles in Fig. 13.2. In these figures, the dipole contributions (indicated by the dashed, blue curves) become small when $a > 2$, and the bubble contributions (indicated by the solid, red curves) are very nearly the same for the parallel and anti-parallel cases. Also, except when $a \sim 1$, the initial push of each bubble is stronger than the subsequent pull, so that the net effect of the bubble is to create a repulsive force between the spheres. This is quite different than the static force between the dipoles, which is repulsive when the moments are parallel and attractive when the moments are anti-parallel.

13.3 A Bubble from Sphere 1 Passing the Dipole Field of Sphere 2

As a final contribution to Chap. 13, we consider the field patterns that are obtained when a bubble, produced by turning the current on suddenly in one sphere, passes the other sphere. We use the same geometry that is shown in Figs. 13.1 and 13.5 with the half-separation distance, a, between the spheres equal to $8R$. Also, we turn the currents on suddenly in both spheres, but focus on the fields around one sphere as the bubble from the other sphere passes it, as shown in Fig. 13.7.

Figure 13.7 shows snapshots of the field geometry on the inbound and outbound passages of the bubble. In the upper panels, the target sphere has a positive dipole moment with its field directed upward in the positive $+z$-direction. In the lower panels, the dipole moment of the target sphere is reversed, now pointing downward. (This is opposite the arrangement in Fig. 13.5 in which the left sphere had a negative dipole moment and the right sphere had a positive moment.) In summary, we are sending a bubble of clockwise circulating flux past a dipole, in one case with its dipole moment pointing upward as shown in the upper panels of Fig. 13.7 and the other case with its moment pointing downward as shown in the lower panels.

In the upper panels, the bubble encounters the exterior field lines of the positive dipole, which are pointed downward at the equator. This means that the bubble's leading field reinforces the field of the dipole and the bubble's trailing field opposes the dipole field. Consequently, on the inbound path, Pinchoff occurs first at the middle of the bubble and continues to the trailing edge where a slightly pinched loop is visible at $r_L = 15.5$. The two lobes of the divided bubble then pass around the sphere. On the outbound path, a stretched equatorial loop is visible at $r_L = 20.15$. It has the same characteristic shape that we have seen in previous chapters where a field of one polarity passes through a static field of opposite polarity, and marks the start of the reconnection that will separate the bubble from the dipole field.

The reverse process is shown in the lower panels where the dipole moment of the target sphere is pointed downward. Consequently, in these lower panels, the Pinchoff occurs in the leading part of the bubble. The familiar, elongated shape of the collapsing loop at the leading edge of the bubble marks the end of the reconnection process around $r_L = 14.25$. Then, after passing the sphere, the pinched loop is visible at $r_L = 18.5$. This loop marks the onset of the reconnection that rejoins the separate lobes into a single bubble that proceeds outward beyond the target sphere. These changes are essentially the same as those observed for the

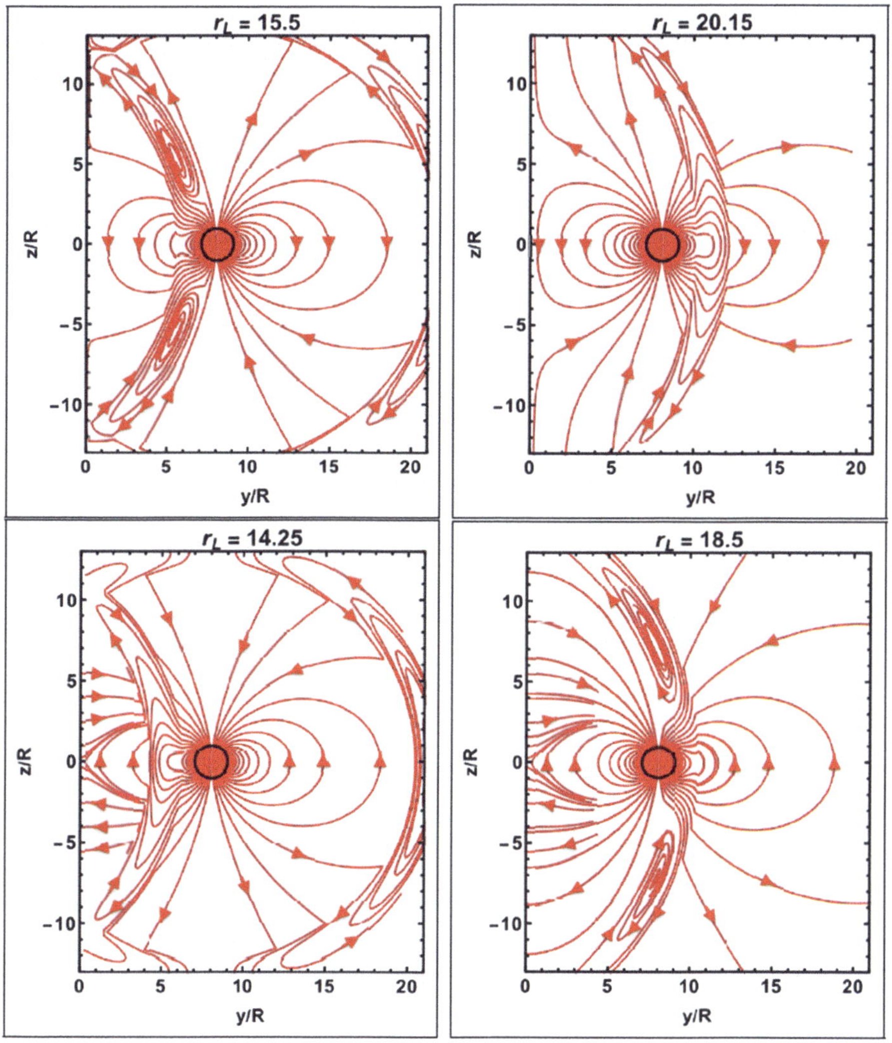

Fig. 13.7 The bubble from the sudden turn-on of the current in a positive dipole is shown passing through the field of a parallel dipole (top panels) and an anti-parallel dipole (bottom panels). In each case, the source sphere is located a distance $2a$ from the target sphere, as shown in Figs. 13.1 and 13.5

parallel configuration in the upper panels when the images are observed in reverse order of time.

The differences between the pinched and elongated shapes of the loops seen on the inbound and outbound paths are due to the oppositely directed gradient of the field in the bubble on those two paths. On the outbound path, the strength of the meridional field is largest at the rear of the bubble and weakens toward the center, tending to match the decreasing strength of the dipole field with distance along the outbound path and causing the loops to stretch out. On the inbound path, the meridional field strength is also largest at the rear of the bubble, but the dipole field strength increases as the bubble approaches the sphere. These oppositely directed gradients shrink the reconnection region and cause the loops to be more compressed. Thus, the inbound-outbound asymmetry of the loops is a consequence of the properties of the bubble from a sudden turn-on. We would not expect such an asymmetry in the sheaths of bubbles produced by a linear current ramp.

We can obtain a more precise explanation of this difference from the equations for the neutral points. On the "equator," the radial component of the field is always zero, so it is only necessary to consider the poloidal component, B_θ, which is

$$\frac{B_\theta}{B_0} = \frac{1}{2} r^{-3} (f_o - r f_o') + \frac{1}{2} \rho^{-3}, \tag{13.29}$$

where the first term is taken from Eq. (3.3b) with r the distance from the remote source sphere to the field point near the target sphere, and $\rho = r - 2a$ on the outbound path and $\rho = 2a - r = -(r - 2a)$ on the inbound path. This difference in the sign of ρ is responsible for the difference in the appearance of the field on the inbound and outbound paths, as we shall see next.

The neutral point is obtained by setting $B_\theta = 0$, which means that

$$f_o(r, r_L) - r f_o'(r, r_L) + \left(\frac{r}{r - 2a} \right)^3 = 0 \tag{13.30}$$

on the outbound path, and that

$$f_o(r, r_L) - r f_o'(r, r_L) - \left(\frac{r}{r - 2a} \right)^3 = 0 \tag{13.31}$$

on the inbound path. For the neutral points to occur in the trailing part of the bubble, they must satisfy these equations when r lies in the range $(r_L - 2, r_L)$.

Figure 13.8 shows these two neutral-point curves plotted in a conventional (r_L, r) space-time map, but without the usual field-line tracks. The half-separation between the spheres is $a = 8$, as used in the previous figures of this chapter. In the left panel, the inbound track progresses asymptotically from the center of the bubble (indicated by the dotted line) to its final location at the rear of the bubble where $r = r_L - 2$ without ever having a vertical slope. The black dot just indicates the endpoint of the curve. In the right panel, the outbound track begins at the rear of the bubble and moves asymptotically toward the center. On this track, the black dot indicates a point (20.296, 18.408) where the slope of the contour is vertical. We can understand this behavior as follows.

If the neutral-point curve has a vertical slope, it will occur where $dr/dr_L = \infty$. Solving equation (13.30) for dr/dr_L, we obtain

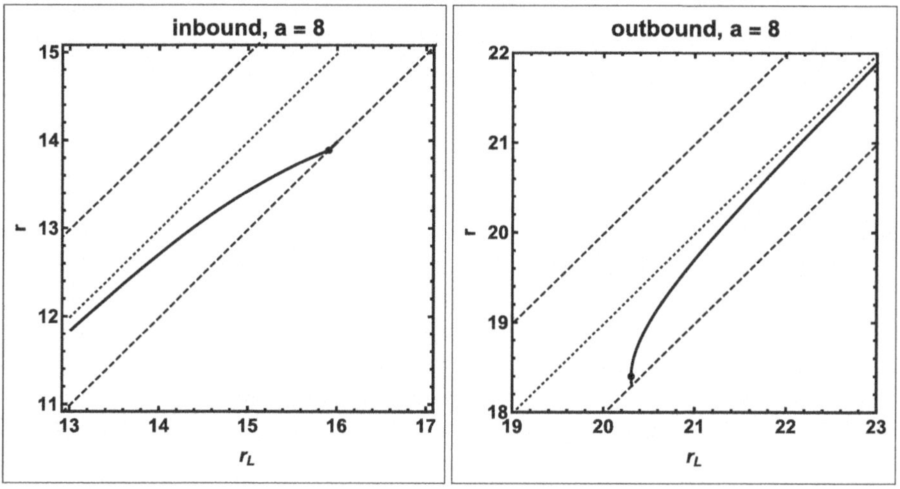

Fig. 13.8 Neutral-point contours for the field shown in the upper panels of Fig. 13.7. The dashed lines indicate the leading and trailing edges of the bubble and the dotted line indicates their center where $r = r_L - 1$. On the inbound path (left), the contour never has a vertical slope. On the outbound path (right), the contour has a vertical slope at the black dot where $(r_L, r) = (20.296, 18.408)$. This accounts for the familiar stretched shape of the first collapsing loop. r is the distance between the field point and the source sphere, so that $y = r - a$ in terms of the coordinates in Fig. 13.7

$$\frac{dr}{dr_L} = \frac{\partial(f_o - rf_o')/\partial r_L}{r\left\{f_o'' + 3(2a)r/(r-2a)^4\right\}} \tag{13.32}$$

on the outbound path. This means that a vertical slope, if it exists, must occur where

$$f_o'' + 3\frac{(2a)r}{(r-2a)^4} = 0. \tag{13.33}$$

This equation is similar to the condition, $\partial^2 f/\partial r^2 = 0$, that we encountered in the discussion following Eq. (6.5) for the linear on-ramp in Chap. 6. However, in Eq. (13.33), there is an extra term due to the gradient of the field of the target dipole. It is this gradient relative to the value of f_o'' that causes the different appearances of the field on the inbound and outbound paths.

Substituting the value of $f_o(r, r_L)$ from Eq. (2.11a), we obtain

$$r_L - 1 = 2r\left\{1 - \frac{2a}{(r-2a)^4}\right\} \tag{13.34}$$

for the outbound path. Then, reversing the sign of the a-dependent term, we obtain

$$r_L - 1 = 2r\left\{1 + \frac{2a}{(r-2a)^4}\right\} \tag{13.35}$$

for the inbound path. Together Eqs. (13.30) and (13.34) have a unique solution for r in the range $(r_L - 2, r_L)$, but Eqs. (13.31) and (13.35) do not have a solution in that range. Therefore, the outbound track of neutral points has a location where the slope is vertical, but the inbound track does not have such a location. This accounts for the familiar stretched shape of the equatorial loop in the right panel of Fig. 13.7 and the lack of this shape in the left panel. A similar argument accounts for the presence of the stretched loop on the inbound path when the dipole moment is reversed, as shown in the lower panels of Fig. 13.7.

If we consider both spheres at once, as in Fig. 13.1 or Fig. 13.5, each bubble will create the same pattern of reconnection as it passes through the dipole field of the other sphere. In the parallel configuration, the reconnection will occur in the leading part of each bubble and the resulting field-line patterns will be those of the upper panels of Fig. 13.7. In the anti-parallel configuration, the reconnection will occur in the trailing part of each bubble and the field-line patterns will be those of the lower panels.

The changes in Fig. 13.7 are similar to those that we observed in Chap. 8 for a plane wave moving past a dipole field in two dimensions. More precisely, the changes are similar to what we would have observed if we had sent a plane wave of opposite polarity directly behind the original plane wave to restore the dipole field to its original configuration after the passage of the wave. However, in the 3-D examples shown in Fig. 13.7, the inbound and outbound solutions are slightly different due to the front-back asymmetry of the field of the bubble from the sudden turn-on. In our 2-D case, the magnetic field of the plane wave was symmetric, as shown in the right panel of Fig. 8.3. This front-back symmetry caused the inbound and outbound configurations to be the same.

I should add that it was more difficult to draw the field lines for the two-sphere configuration than for the one-sphere sources described in other chapters. This was due to the lack of spherical symmetry and the consequent fact that the field lines were no longer contours of constant flux. Therefore, it was necessary to return to conventional field-line equations to draw the field lines. To avoid this task, I used the Mathematica© "StreamPlot" procedure to draw field lines that originated in specific points. For most of the field lines, I chose the starting points to be spaced sinusoidally along the equatorial diameter of each sphere. However, it was necessary to select additional points to fill in some of the empty spaces. This was something of an art, but it seems to have achieved the desired results.

Chapter 14
Concentric Sources

In the previous chapter, we considered the interaction between two spherical current sources separated by a distance $2a$. In general, the lack of spherical symmetry made it difficult to obtain analytical solutions. However, by decreasing the size of one of those spheres and placing it at the center of the other, we obtain a symmetric configuration which permits analytical solutions to the kinds of problems encountered for the asymmetric configuration.

Let the radius of the inner sphere be R_i and the much larger radius of the outer sphere be R_o. Suppose that the current in the inner sphere has not been turned on yet, but that the current in the outer sphere was turned on so long ago that its transient has left the outer sphere. In this case, there is a uniform field, B_o, inside the radius, R_o, of the outer sphere and a corresponding dipole field outside. Also, at this time, the outer sphere has provided a total energy, $3\mathcal{E}_o$, of which $\mathcal{E}_o$ lies in the uniform field, B_o, and $2\mathcal{E}_o$ is distributed between the exterior dipole field and the bubble, which end up with $(1/2)\mathcal{E}_o$ and $(3/2)\mathcal{E}_o$, respectively. Following our convention, $\mathcal{E}_o = (\frac{4}{3}\pi R_o^3)(B_o^2/2\mu_0)$.

Next, suppose that the current in the inner sphere is turned on suddenly. Because $R_o \gg 3R_i$, the trailing edge of the bubble will leave the inner sphere at the time $t = 3R_i/c$, well before its leading edge reaches the outer sphere. But as time increases, this bubble eventually passes through the outer sphere and follows its predecessor into the distance carrying an energy of $(3/2)\mathcal{E}_i$, where $\mathcal{E}_i = (\frac{4}{3}\pi R_i^3)(B_i^2/2\mu_0)$. So the total radiated energy is $\mathcal{E}_{rad} = (3/2)\mathcal{E}_i + (3/2)\mathcal{E}_o$. But how much energy is left in the composite field?

To calculate this field energy, we divide the space into three regions: region $\mathcal{R}_I$ inside the inner sphere where $r < R_i$, region $\mathcal{R}_{II}$ between the two spheres where

---

Electronic Supplementary Material The online version of this chapter (https://doi.org/10.1007/978-3-030-40264-8_14) contains supplementary material, which is available to authorized users.

N. R. Sheeley, Jr., *Transient Magnetic Fields*,
https://doi.org/10.1007/978-3-030-40264-8_14

$R_i < r < R_o$, and region $\mathcal{R}_{III}$ outside the outer sphere where $r > R_o$. The energy in $\mathcal{R}_I$ is simply

$$\mathcal{E}_I = \left(\frac{4}{3}\pi R_i^3\right)\frac{(B_i + B_o)^2}{2\mu_0} = \frac{4}{3}\pi\left\{R_i^3 B_i^2 + 2R_i^3 B_i B_o + R_i^3 B_o^2\right\}\left(\frac{1}{2\mu_0}\right). \tag{14.1}$$

The energy in $\mathcal{R}_{II}$ is obtained by combining the components of the dipole and uniform field before squaring and integrating as follows:

$$\mathcal{E}_{II} = \int_0^{2\pi} d\phi \int_0^{\pi} \sin\theta d\theta \int_{R_i}^{R_o} r^2 dr \left[\left\{B_i\left(\frac{r}{R_i}\right)^{-3} + B_o\right\}^2 \cos^2\theta \right.$$
$$\left. + \left\{\frac{1}{2}B_i\left(\frac{r}{R_i}\right)^{-3} - B_o\right\}^2 \sin^2\theta\right]\left(\frac{1}{2\mu_0}\right). \tag{14.2}$$

After performing the integration, we obtain:

$$\mathcal{E}_{II} = \frac{2}{3}\pi\left\{R_i^3 B_i^2 - \frac{Ri^6}{R_o^3}B_i^2\right\}\left(\frac{1}{2\mu_0}\right) + \frac{4}{3}\pi\left\{R_o^3 B_o^2 - R_i^3 B_o^2\right\}\left(\frac{1}{2\mu_0}\right). \tag{14.3}$$

The energy in $\mathcal{R}_{III}$ is obtained by combining the two dipole fields before squaring and integrating:

$$\mathcal{E}_{III} = \int_0^{2\pi} d\phi \int_0^{\pi} \sin\theta d\theta \int_{R_i}^{R_o} r^2 dr \left[\left\{(B_i + B_o)\left(\frac{r}{R_i}\right)^{-3}\right\}^2 \cos^2\theta \right.$$
$$\left. + \left\{\frac{1}{2}(B_i + B_o)\left(\frac{r}{R_i}\right)^{-3}\right\}^2 \sin^2\theta\right]\left(\frac{1}{2\mu_0}\right). \tag{14.4}$$

Again, performing the integration, we obtain:

$$\mathcal{E}_{III} = \frac{2}{3}\pi\left\{\frac{Ri^6}{R_o^3}B_i^2 + 2R_i^3 B_i B_o + R_o^3 B_o^2\right\}\left(\frac{1}{2\mu_0}\right). \tag{14.5}$$

Finally, combining Eqs. (14.1), (14.3), and (14.5), we obtain the volume energy, $\mathcal{E}_{vol}$, in the three regions:

$$\mathcal{E}_{vol} = \frac{3}{2}\left(\frac{4}{3}\pi R_i^3\right)\frac{B_i^2}{2\mu_0} + \frac{3}{2}\left(\frac{4}{3}\pi R_o^3\right)\frac{B_o^2}{2\mu_0} + 3\left(\frac{4}{3}\pi R_i^3\right)\frac{B_i B_o}{2\mu_0}. \tag{14.6}$$

Expressed another way,

$$\mathcal{E}_{vol} = \frac{3}{2}\,\mathcal{E}_i + \frac{3}{2}\,\mathcal{E}_o + 3\left(\frac{4}{3}\pi R_i^3\right)\frac{B_i B_o}{2\mu_0}. \tag{14.7}$$

When this volume energy of the composite field is combined with the energy carried away by the outgoing bubbles, one obtains the total energy provided by the two spheres:

$$\mathcal{E}_{tot} = \mathcal{E}_{rad} + \mathcal{E}_{vol} = 3\,\mathcal{E}_i + 3\,\mathcal{E}_o + 3\left(\frac{4}{3}\pi R_i^3\right)\frac{B_i B_o}{2\mu_0}. \tag{14.8}$$

In Chap. 4, we saw that the total cost of releasing a bubble is 3 units of energy. Thus, acting separately, the cost of releasing the bubbles would have been $3\mathcal{E}_i + 3\mathcal{E}_o$. But when acting together, the two spheres leave an extra energy $3(\frac{4}{3}\pi R_i^3)(B_i B_o/2\mu_0)$ in the composite field. Where did this extra energy come from? The second bubble must have left it in the combined field. To end up with a radiated energy of $(3/2)\mathcal{E}_i$, the second bubble must have acquired the balance as it passed the current in the outer sphere.

To confirm this, we calculate the energy transfer as the second bubble passed the current in the outer sphere. Referring to Eq. (4.5), the energy given up by the current is $-J_\phi E_\phi$, integrated over a volume enclosing the current and a time corresponding to the passage of the bubble:

$$\Delta\mathcal{E} = \int\int(-J_\phi E_\phi)dt\,dV. \tag{14.9}$$

In this case, we take J_ϕ from Eq. (2.1) with $R = R_o$ and $B_0 = B_o$. Also, we take E_ϕ from Eq. (3.3c) with r in units of R_i, and we express time in units of R_i/c to obtain

$$\Delta\mathcal{E} = -\int_0^{2\pi} d\phi \int_0^{\pi} \sin\theta d\theta \int_{R_o-\epsilon}^{R_o+\epsilon} r^2 dr \int_{\rho}^{\rho+2} dr_L\left(\frac{R_i}{c}\right)$$
$$\times\left\{\frac{3}{2}\frac{B_o}{\mu_0}\delta(r - R_o)\sin\theta\right\}\left\{-\frac{1}{2}cB_i R_i^2 r^{-2}\frac{\partial f}{\partial r_L}\sin\theta\right\}, \tag{14.10}$$

where ϵ is any small positive number that makes the r-integral enclose the sphere, and $\rho = R_o/R_i$. The r_L-integral gives 1, and the Dirac delta function forces the r-integral to be 1. The ϕ-integral is 2π and the θ-integral is 4/3. Combining these factors, we get

$$\Delta\mathcal{E} = +3\left(\frac{4}{3}\pi R_i^3\right)\frac{B_i B_o}{2\mu_0}, \tag{14.11}$$

which is just the amount needed to supply the extra energy in Eq. (14.8).

It is interesting to compare this extra energy with the dipole moment of the inner sphere. For this purpose, we regard the spherical current as a collection of infinitesimal current loops, each with a dipole moment $A(\theta)di$. Thus, the dipole moment of the spherical current is

$$\begin{aligned} M_i &= \int_{R_i-\epsilon}^{R_i+\epsilon} rdr \int_0^{\pi} d\theta\, J_\phi \pi (r\sin\theta)^2 \\ &= \int_{R_i-\epsilon}^{R_i+\epsilon} rdr \int_0^{\pi} d\theta \left\{ \frac{3}{2}\frac{B_i}{\mu_0}\delta(r-R_i)\sin\theta \right\} \pi (r\sin\theta)^2, \end{aligned} \tag{14.12}$$

where J_ϕ was obtained from Eq. (2.1) with $R = R_i$ and $B_0 = B_i$. Here, the r-integral gives R_i^3 and the θ-integral again gives 4/3. Combining these factors, we obtain

$$M_i = 3\left(\frac{4}{3}\pi R_i^3\right)\frac{B_i}{2\mu_0}. \tag{14.13}$$

Consequently, $\Delta\mathcal{E} = M_i B_o$ and the total energy provided by the two spherical current sources is

$$\mathcal{E}_{tot} = 3\mathcal{E}_i + 3\mathcal{E}_o + M_i B_o. \tag{14.14}$$

Here, I emphasize that the extra energy is positive when $B_i > 0$ and $B_o > 0$ (i.e., when the dipoles are parallel). The extra energy is positive because the parallel dipoles reinforce each other when one current loop is placed inside the other. The extra energy would be negative if the dipoles were anti-parallel.

Figure 14.1 shows the field lines in the yz-plane at the time $r_L = 8.48$ for $B_o = +1/54$ (left panel) and for $B_o = -1/54$ (right panel). In each case, $B_i = +1$, $R_i = 1$, and $R_o > 14$. The reason for choosing $|B_o|$ so small is to keep the bubble visible for a while before its strength falls below this constant "background" value. This procedure worked well in the left panel, where the bubble is visible moving away from the "equatorial" neutral points (corresponding to the ring-shaped neutral line in three dimensions) that it left in the static field. These neutral points separate the equatorial loops from the open field lines that now link the polar regions of the sphere to the uniform background field, $B_o\mathbf{e_z}$. These neutral points occur at the radial distance $r = (2B_o)^{-1/3}$, which is $r = 3R_i$ when $B_o = 1/54$, as one can see in the left panel. As the bubble continues outward into the background field, flux is peeled away from its leading edge. In this process, the O-type neutral point within the bubble works its way forward to the leading edge where $r = r_L$. Together, the conditions $B_\theta = (1/2)B_i R_i^3 r^{-3}(f - rf') - B_o = 0$ and $r = r_L$ give $r = r_L = (3/4)/B_o$ for the demise of this neutral point, which is $r = r_L = 40.5$ for $B_o = 1/54$. So, if R_o were greater than 40.5, all of the bubble's flux would have been peeled away by the time it arrived there.

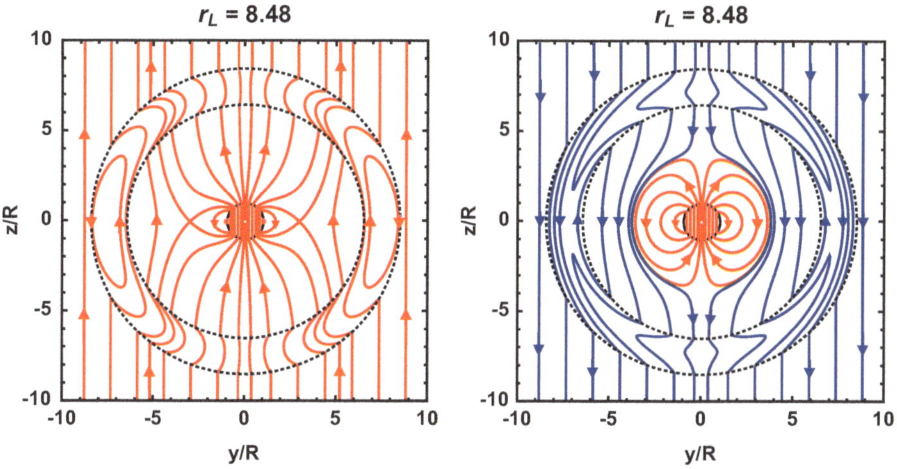

Fig. 14.1 *yz*-views of the evolving field lines for spherical current loops of radius $R_i = 1$, centered in the fields of much larger current loops of radius $R_o > 14$ that were turned on long ago. These snapshots show the results at the elapsed time, $r_L = 8.48$, after the currents on the inner spheres were suddenly turned on. When the dipole moments of the inner and outer spheres are parallel (left panel), the polar field of the inner sphere becomes linked to the uniform field of the outer sphere, but when the dipole moments are anti-parallel (right panel), the field of the inner sphere becomes isolated from the external uniform field

For the anti-parallel configuration in the right panel of Fig. 14.1, the red contours of the outgoing bubble from the inner sphere are no longer visible between the dashed wave fronts by the time that $r_L = 8.48$. Also, at this time, the closed field lines of the inner sphere have become isolated from the surrounding field and are confined to a sphere of radius, $r = (B_o)^{-1/3}$, which is $3(2)^{1/3} = 3.78R_i$ for $B_o = 1/54$, as one can easily confirm by setting $B_r(r, 0) = B_i R_i^3 r^{-3} - B_o = 0$. In general, this radius, $r = R_i(B_i/B_o)^{1/3}$, can lie in the interval (R_i, R_o) only if $B_i > B_o$ and $M_o > M_i$, where $M_o = 3(4\pi R_o^3/3)(B_o/2\mu_0)$ and $M_i = 3(4\pi R_i^3/3)(B_i/2\mu_0)$ are the dipole moments of the currents on each sphere.

Let us summarize the results of placing one spherical current source inside the current on a larger sphere. For the parallel configuration, the polar flux of the inner sphere becomes linked to the uniform field of the outer sphere and the energy of this composite field exceeds that of the separate fields by the amount $\Delta\mathcal{E} = +M_i B_o$. As we showed above, the bubble that left this extra energy in the composite field must be repaid as it passes the current in the outer sphere and departs the region carrying its quota of radiated energy, $(3/2)\mathcal{E}_i$. For the anti-parallel configuration, the field of the inner sphere becomes isolated from the surrounding uniform field of the outer sphere, and this configuration has less energy than the separate fields by the amount $\Delta\mathcal{E} = -M_i B_o$. Consequently, the outgoing bubble has too much energy and must return this excess when it passes the current in the outer sphere. As we shall see next, this energy budget is reversed when the currents are located side-by-side, as they were in Chap. 13.

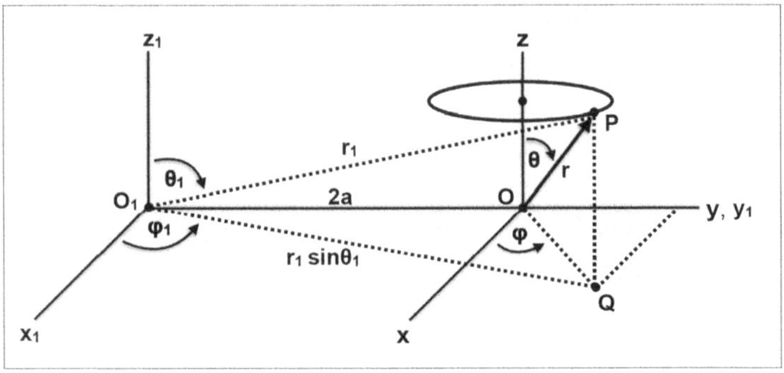

Fig. 14.2 Sketch of geometry used to derive the energy exchange at the second sphere

Figure 14.2 shows the geometry for two identical spherical currents separated by the distance $2a$. We assume that the dipoles are parallel and ask how much energy is acquired as the bubble from sphere 1 (centered at point O_1) passes the current on sphere 2 (centered at point O). We regard the current on the surface of sphere 2 to be composed of a collection of infinitesimal current loops, one of which is shown in the figure. P is a point on that loop, having the spherical coordinates (r, θ, ϕ) with respect to point O and (r_1, θ_1, ϕ_1) with respect to point O_1. Equation (14.9) gives the energy that the current at point P provides to the bubble from sphere 1. In this case, the electric field, $\mathbf{E}$, and the current, $\mathbf{J}$, at point P are given by

$$\mathbf{E} = -\frac{1}{2} c B_0 r_1^{-2} \frac{\partial f}{\partial r_L} \sin\theta_1 \, \mathbf{e}_{\phi_1} \tag{14.15}$$

and

$$\mathbf{J} = \frac{3}{2} \frac{B_0}{\mu_0} \delta(r - R) \sin\theta \, \mathbf{e}_{\phi}, \tag{14.16}$$

where B_0 is the uniform field as defined in Eq. (2.1) and R is the radius of each sphere.

Substituting these values into Eq. (14.9) with $dV = r^2 \sin\theta d\theta d\phi dr$ and $dt = (R/c) dr_L$, we obtain

$$\Delta\mathcal{E} = +\frac{3}{2} \frac{B_0^2}{2\mu_0} R \int r^2 dr \int_0^{2\pi} d\phi \int_0^{\pi} \sin\theta d\theta$$
$$\times \int dr_L \, r_1^{-3} (r_1 \sin\theta_1)(\mathbf{e}_{\phi} \cdot \mathbf{e}_{\phi_1}) \, \delta(r - R) \sin\theta \frac{\partial f}{\partial r_L}. \tag{14.17}$$

Recognizing that $\int dr_L (\partial f / \partial r_L) = 1$ and that the Dirac delta function requires us to replace r by R, we reduce Eq. (14.17) to

$$\Delta\mathcal{E} = +\frac{3}{2}\frac{B_0^2}{2\mu_0}R^3 \int_0^{2\pi} d\phi \int_0^{\pi} \sin\theta^2 d\theta \, r_1^{-3}(r_1 \sin\theta_1)(\mathbf{e}_\phi \cdot \mathbf{e}_{\phi_1}). \tag{14.18}$$

Next, referring to Fig. 14.2, we write

$$r_1 \sin\theta_1 (\mathbf{e}_\phi \cdot \mathbf{e}_{\phi_1}) = r_1 \sin\theta_1 \cos(\phi_1 - \phi) = x\cos\phi + (2a+y)\sin\phi = r\sin\theta + 2a\sin\phi. \tag{14.19}$$

Setting $r = R = 1$ in Eq. (14.19), and substituting the normalized result back into Eq. (14.18), we obtain

$$\Delta\mathcal{E} = +\frac{3}{2}\frac{B_0^2}{2\mu_0}R^3 \int_0^{2\pi} d\phi \int_0^{\pi} \sin^2\theta d\theta \frac{(\sin\theta + 2a\sin\phi)}{(1 + 4a\sin\theta\sin\phi + 4a^2)^{3/2}}, \tag{14.20}$$

where a is now expressed in units of R. Next, we assume that $a \gg 1$ and expand the integrand of Eq. (14.20) in powers of $1/a$, obtaining

$$\Delta\mathcal{E} \approx +\frac{3}{2}\frac{B_0^2}{2\mu_0}R^3 \int_0^{2\pi} d\phi \int_0^{\pi} \sin^2\theta d\theta \times \left\{ \left(\frac{1}{4a^2}\right)\sin\phi + \left(\frac{1}{8a^3}\right)\sin\theta\,(1 - 3\sin^2\phi) + \ldots \right\}. \tag{14.21}$$

The first term in the braces does not contribute to the integral, and the second term gives

$$\Delta\mathcal{E} \approx -\frac{3}{16}\left(\frac{1}{a^3}\right)\left(\frac{4}{3}\pi R^3\right)\left(\frac{B_0^2}{2\mu_0}\right). \tag{14.22}$$

Replacing a by its unnormalized value, a/R, and regrouping the factors of $\Delta\mathcal{E}$, we obtain

$$\Delta\mathcal{E} \approx -\left\{3\left(\frac{4}{3}\pi R^3\right)\left(\frac{B_0}{2\mu_0}\right)\right\}\left\{\frac{1}{2}B_0\left(\frac{R}{2a}\right)^3\right\} = -M_0 B_1, \tag{14.23}$$

where $M_0 = 3(\frac{4}{3}\pi R^3)(B_0/2\mu_0)$ is the dipole moment of the spherical current loop, and $B_1 = (1/2)B_0(R/2a)^3$ is the θ-component of the field that sphere 1 produces at the location of sphere 2. Because this is the energy that the current in sphere 2 gives up when the bubble from sphere 1 passes it, the negative value means that the bubble actually returns the energy, $M_0 B_1$, to the sphere. The reason that the bubble had this excess energy is that it left too little energy in the composite field. Consequently, the composite field of the parallel dipoles has less energy than the combined energy of the separate fields.

At this point, it would be tempting to say that the other bubble returned the same energy and that the total deficit should be twice this amount. But that would be

Fig. 14.3 Excess field energies for dipoles with their moments parallel or anti-parallel. Parallel (||) dipoles reinforce each other when one is placed inside the current system of the other, but oppose each other when placed outside. The excess energies are reversed for anti-parallel dipoles

	$\parallel$	$\nparallel$
inside	$\Delta\mathcal{E} = + M_i B_o$	$\Delta\mathcal{E} = - M_i B_o$
outside	$\Delta\mathcal{E} = - M_0 B_1$	$\Delta\mathcal{E} = + M_0 B_1$

wrong! The energy budget of the other bubble has already been included in the $3\mathcal{E}_0$ required to turn its current source on. We could have turned that current on long ago and waited until the transient left the region before turning on the current in the second sphere, just as we did with the concentric spheres. The correct answer is that the total energy required to set up the composite field of the parallel dipoles is

$$\mathcal{E}_{tot} = 3\mathcal{E}_0 + 3\mathcal{E}_0 - M_0 B_1. \tag{14.24}$$

For the anti-parallel dipoles of Fig. 13.5, the extra energy would be $+M_0 B_1$.

Figure 14.3 summarizes the results for the four cases we have considered: parallel and anti-parallel dipoles with one sphere placed inside the other, and parallel and anti-parallel dipoles placed side-by-side at a great distance from each other. It is instructive to analyze these results both mathematically and in terms of the associated field geometry. Mathematically, the extra energy, $\Delta\mathcal{E}$, can be represented by the term $+\mathbf{M}\cdot\mathbf{B}$, where $\mathbf{M} = \mathbf{M_i}$ and $\mathbf{B} = \mathbf{B_o}$ for the concentric spheres and $\mathbf{M} = \mathbf{M_0}$ and $\mathbf{B} = \mathbf{B_1}$ for the spheres that are side-by-side. Thus, for parallel, concentric dipoles, $\mathbf{M_i}$ and $\mathbf{B_o}$ are parallel and $\Delta\mathcal{E} = +M_i B_o$. For the anti-parallel, concentric dipoles, $\mathbf{M_i}$ and $\mathbf{B_o}$ are anti-parallel so that $\Delta\mathcal{E} = -M_i B_o$. For the exterior, side-by-side dipoles, the upward-directed field from one sphere bends over and points downward at the other sphere, so that $\mathbf{M}_0\cdot\mathbf{B_1} = -M_0 B_1$ for the parallel configuration and $\mathbf{M}_0\cdot\mathbf{B_1} = +M_0 B_1$ for the anti-parallel configuration.

All of these values are consistent with the detailed derivations above as well as with general expressions in terms of the dot products, either $+\mathbf{M}_i\cdot\mathbf{B}_o$ for the inside configuration or $\Delta\mathcal{E} = +\mathbf{M}_0\cdot\mathbf{B}_1$ outside. Note, in particular, that the extra energies are given by $+\mathbf{M}\cdot\mathbf{B}$, and not $-\mathbf{M}\cdot\mathbf{B}$, as one might have expected for a dipole placed in an external magnetic field. Thus, the energy that we obtained by letting the flattened fields of Fig. 13.1 push apart is $(3/16)a^{-3}\ \mathcal{E}_0$, which agrees with the magnitude of $-M_0 B_1$ in Eq. (14.23), but not with its sign. This paradox has been discussed by Feynman in his Lectures on Physics (Feynman et al. 1964), and occurs because we are considering all of the energy and not just the mechanical energy of the dipoles.

The analysis in terms of the field geometry is even more interesting because it leaves us with a mental picture of the results. As shown in Fig. 14.1, when the current in a small sphere is suddenly turned on, the polar fields of that sphere become linked to the uniform field of the larger sphere if the dipole moments of the two spheres are parallel. If the dipole moments are anti-parallel, the field of the inner sphere becomes isolated from the field of the larger sphere whose field lines bend smoothly around it. Thus, the linked field has extra energy and the isolated field has less energy than the amount of energy that would be contained in the separate fields.

In Fig. 13.1, the field lines of the parallel dipoles became flattened against each other at the mid-plane between the two spheres. As revealed in Eq. (14.24), this configuration has less energy than is contained in the separate fields, and is analogous to the isolated field in the right panel of Fig. 14.1. In Fig. 13.5, reconnection at the mid-plane caused the polar fields of the anti-parallel dipoles to be linked together, producing a configuration similar to the linked parallel fields of the concentric dipoles in the left panel of Fig. 14.1. Thus, there is a general principle that applies to all of these spherical current systems, regardless of whether one lies inside the other or whether they lie side-by-side: It takes more energy to produce a linked configuration (and less energy to produce an isolated configuration) than it takes to create the separate fields.

Next, we return to the concentric spheres and ask what would happen if the current on the inner sphere were turned on first, and then, after a long time, the current in the outer sphere were turned on. The energy required to turn on the current in the inner sphere is $3\mathcal{E}_i = 3(\frac{4}{3}\pi R_i^3)(B_i^2/2\mu_0)$, of which $(3/2)\mathcal{E}_i$ has escaped in the bubble and $(3/2)\mathcal{E}_i$ is stored in the field. Now, suddenly turn on the current in the outer sphere. If the inner and outer dipoles are parallel, then a wave would move inward, increasing the strength of the uniform field inside the inner sphere to the value $B_i + B_o$. The wave would then move back outward from the center, leaving a "transition" field in the region between the two spheres and a composite dipole of strength $B_o + B_i(R_i/R_o)^3$ outside the outer sphere. The total energy required to produce this field would be the same that we found when we turned the outer current on first, namely $3\mathcal{E}_i + 3\mathcal{E}_o + M_i B_o$.

In this case, the extra energy must have been put there by the bubble from the outer sphere, which must have acquired this energy as it passed the current on the inner sphere. Consequently, we expect $\int\int(-\mathbf{J}\cdot\mathbf{E})dVdt$, as evaluated at the inner sphere, to equal the extra energy $M_i B_o$. This equality is easily verified by returning to Eq. (14.10) and interchanging the inside (i) and outside (o) terms and replacing $\int_\rho^{\rho+2} dr_L$ by $\int_{2-\rho}^{2+\rho} dr_L$ with $\rho \rightarrow R_i/R_o$. In this case, $\int_{2-\rho}^{2+\rho} dr_L(\partial f/\partial r_L) = f(\rho, 2+\rho) - f(\rho, 2-\rho) = \rho^3 = (R_i/R_o)^3$. Therefore,

$$\Delta\mathcal{E} = +3\left(\frac{4}{3}\pi R_o^3\right)\left(\frac{B_i B_o}{2\mu_0}\right)(R_i/R_o)^3 = +3\left(\frac{4}{3}\pi R_i^3\right)\left(\frac{B_i B_o}{2\mu_0}\right) = M_i B_o, \tag{14.25}$$

as we expected. The only difference between this result and the previous one is that the outer sphere provides the extra energy when it is turned on first. In each case, the second bubble transfers the energy, which is $M_i B_o$ in both cases and never $M_o B_i$,

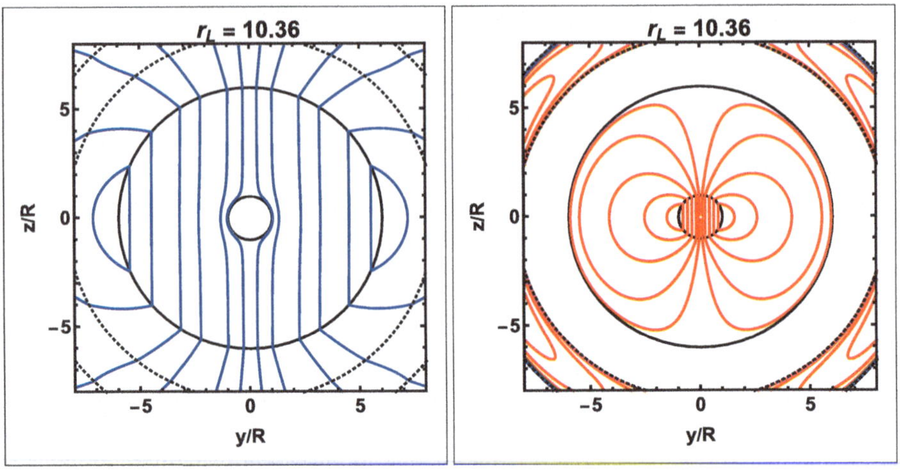

Fig. 14.4 Field lines of anti-parallel, concentric dipoles when $r_L = 10.36$ units of R_i/c after the current on the inner sphere was turned on. In both panels, the bubble from the inner sphere lies between the dotted circles. In the left panel, the dipole field strengths are $B_i = -B_o$, which causes the central field to vanish, and in the right panel, $B_i = -B_o(R_o/R_i)^3$, which causes the outer dipole field to vanish. These are the limiting cases in which the "conducting" surface between the concentric dipoles lies at the extremes of R_i and R_o. The yz-coordinates are plotted in units of $R = R_i$

which is larger than $M_i B_o$ by a factor of $(R_o/R_i)^3$. The same principle applies when the dipoles are anti-parallel and the composite field has less energy.

Figure 14.4 shows special cases of anti-parallel dipoles for which the composite field vanishes inside the inner sphere and outside the outer sphere, respectively. In both cases, the current on the outer sphere was turned on first. Then, after sufficient time for the first bubble to escape from the system, the current in the inner sphere is turned on suddenly at time $r_L = 0$. Each panel shows the resulting field at $r_L = 10.36$ (units of R_i/c) when the bubble from the inner sphere lies between the two dotted circles just leaving field of view.

In the left panel, $B_o = -B_i$, so that the combined field inside the inner sphere is 0. The magnetic field lines sweep smoothly around the sphere, as if it were a superconductor. Outside the outer sphere, the fields combine to give a resultant dipole field whose strength is $B_o\{1-(R_i/R_o)^3\} \approx B_o$. Also, referring to Eq. (14.11), the composite field involves an energy loss $\Delta\mathcal{E} = 3(\frac{4}{3}\pi R_i^3)(B_i^2/2\mu_0) = 3\mathcal{E}_i$. This means that the total energy required to create the composite field is $\mathcal{E}_{tot} = 3\mathcal{E}_i + 3\mathcal{E}_o - 3\mathcal{E}_i = 3\mathcal{E}_o$, which is the energy required to create the field of the outer sphere and its bubble when the inner sphere is not present.

In the right panel, $B_o = -B_i(R_i/R_o)^3$, which makes the combined field outside the outer sphere vanish while leaving the uniform field inside the inner sphere with the strength $B_i\{1-(R_i/R_o)^3\} \approx B_i$. Referring again to Eq. (14.11), we can evaluate the energy loss for this composite field. Substituting $B_i = -B_o(R_o/R_i)^3$

into Eq. (14.11) gives an energy decrease of $\Delta\mathcal{E} = 3(\frac{4}{3}\pi R_o^3)(B_o^2/2\mu_0) = 3\mathcal{E}_o$. Consequently, the total energy required to create the composite field is $\mathcal{E}_{tot} = 3\mathcal{E}_i + 3\mathcal{E}_o - 3\mathcal{E}_o = 3\mathcal{E}_i$, which is the energy required to create the field of the inner sphere and its bubble when the outer sphere is not present.

To summarize both cases, when we adjust the field strengths to make the inner field vanish, the result is an exterior dipole field of approximate strength B_o and a total energy "cost" of $3\mathcal{E}_o$; when we make the outer field vanish, the result is a uniform field inside the inner sphere with an approximate strength of B_i at a total cost of $3\mathcal{E}_i$. In each case, we have isolated a spherical region. In the left panel, it is a vacuum of radius, R_i, trapped at the center of a dipole field of approximate strength, B_o. In the right panel, it is a dipole field of approximate strength, B_i, trapped in the sphere of radius, R_o, with nothing outside. The vacuum is inside the outer dipole and outside the inner dipole.

Note that in each of these cases, the energy transferred by the second bubble, $\Delta\mathcal{E}$, is much less than the total field energy, $\mathcal{E}_{tot}$. This follows from the relation $\mathcal{E}_i/\mathcal{E}_o = (R_i/R_o)^3(B_i/B_o)^2$. Thus, if $B_i = -B_o$, then $\mathcal{E}_i/\mathcal{E}_o = (R_i/R_o)^3 \ll 1$, and therefore $\Delta\mathcal{E}/\mathcal{E}_{tot} = 3\mathcal{E}_i/3\mathcal{E}_o = (R_i/R_o)^3 \ll 1$. However, if $B_o = -B_i(R_i/R_o)^3$, then $\mathcal{E}_o/\mathcal{E}_i = (R_i/R_o)^3 \ll 1$, and therefore $\Delta\mathcal{E}/\mathcal{E}_{tot} = 3\mathcal{E}_o/3\mathcal{E}_i = (R_i/R_o)^3 \ll 1$. Thus, in each of these cases, $\Delta\mathcal{E}/\mathcal{E}_{tot} = (R_i/R_o)^3$, so that the energy deficit relative to the total energy in the composite field (plus the two bubbles) is a measure of the relative volumes of the two spheres. In particular, this deficit ratio is small when the spheres differ greatly in size.

Finally, it is interesting to consider what would happen if one of the currents were repeatedly turned on and off. For simplicity, consider the example illustrated in the right panel of Fig. 14.4, in which the field outside the outer sphere vanished. In this case, the current in the outer sphere was turned on first and left on. Consequently, it will be the beneficiary of the energy, $3\mathcal{E}_o$, that is exchanged each time that the current on the inner sphere is turned on or off, gaining energy during the turn-on and losing it during the turn-off. Thus, the net energy emitted by the outer current alternates between 0 and $3\mathcal{E}_o$. The inner current will give up energy during the turn-on ($3\mathcal{E}_i/2$ for the bubble and $3\mathcal{E}_i/2$ for the field), but not during the turn-off when the bubble acquires its energy from the field. The result will be a series of "on" and "off" bubbles emitted by the inner current, alternately gaining and losing energy, $3\mathcal{E}_o$, from the field and transferring it to the outer current before carrying away $3\mathcal{E}_i/2$ in radiation. The resulting field would oscillate between the isolated dipole configuration (in the right panel of Fig. 14.4) and the field of a uniformly magnetized outer sphere.

Chapter 15
The Boundary-Value Approach

In the previous chapters, we derived the electric and magnetic fields from the known properties of the electric currents that generated them. In this chapter, we assume that the radial component of magnetic field on the surface of the sphere is known as a function of time, and we use this boundary condition to select the correct solution of the wave equation for the fields.

For simplicity, we again consider an axisymmetric field and work with the vector potential which has only a single component, $A_\phi(r, \theta, t)$, that satisfies the wave equation in spherical coordinates as given in Eq. (2.12). The standard solution to this equation (Jackson 1975) has terms of the form

$$A_\phi(r, \theta, t) \sim h_\ell(kr) \frac{\partial P_\ell(\cos\theta)}{\partial\theta} e^{\pm ikct}, \tag{15.1}$$

where the integer, ℓ, and the wave number, k, are constants of integration, P_ℓ is a Legendre polynomial of order ℓ, and $h_\ell(kr)$ is a Hankel function satisfying

$$h_\ell'' + \frac{2}{r} h_\ell' + \left\{k^2 - \frac{\ell(\ell+1)}{r^2}\right\} h_\ell = 0. \tag{15.2}$$

In general, we would construct A_ϕ by summing over ℓ and integrating over k. However, for the dipole field, we need to consider only the term with $\ell = 1$. This means that we must use the radial function, $h_1(kr)$, defined by

$$h_1(kr) = \left\{-\frac{i}{(kr)^2} - \frac{1}{kr}\right\} e^{ikr}, \tag{15.3}$$

and its complex conjugate. We can express A_ϕ in terms of functions with real exponents by making the substitution $k \rightarrow ik$ (where $i = \sqrt{-1}$) and integrating over real k as follows:

N. R. Sheeley, Jr., *Transient Magnetic Fields*,
https://doi.org/10.1007/978-3-030-40264-8_15

$$A_{\phi}^{out}(r,\theta,r_L) = \left(\frac{B_0R}{2}\right) r^{-2} \sin\theta \int_0^{\infty} A^{out}(k)dk(1-kr)e^{k(r-r_L)}, \tag{15.4}$$

and

$$A_{\phi}^{in}(r,\theta,r_L) = \left(\frac{B_0R}{2}\right) r^{-2} \sin\theta \int_0^{\infty} A^{in}(k)dk(1+kr)e^{-k(r+r_L)}. \tag{15.5}$$

Here, I have combined the time-dependent and space-dependent factors to give the outgoing and ingoing waves, A_{ϕ}^{out} and A_{ϕ}^{in}, respectively. Distance, r, is expressed in units of R, and r_L refers to time in units of R/c. Also, I introduced the factor, $B_0R/2$, so that these equations have the same form as Eq. (2.9). In this case, the integrals in Eqs. (15.4) and (15.5) are the f-values for the outgoing and ingoing waves, and both of them satisfy the wave equation in the form given by Eq. (2.13).

So far, our only assumption has been that the field at $r = 1$ has a dipole configuration. Next, our objective is to determine the field everywhere from a knowledge of B_r on the surface of the sphere, or more precisely, from $f(1, r_L)$ in the equations $B_r(1,\theta,r_L) = B_0 f(1,r_L)\cos\theta$ and $A_{\phi}(1,\theta,r_L) = (B_0R/2) f(1,r_L)\sin\theta$. The integrals in Eqs. (15.4) and (15.5) provide the means of doing this. At $r = 1$, they have the common value, $f(1, r_L)$, which is also the magnetic flux (in units of Φ_0) crossing the sphere as a function of time:

$$f(1,r_L) = \int_0^{\infty} A^{out}(k)(1-k)e^{k(1-r_L)}dk = \int_0^{\infty} A^{in}(k)(1+k)e^{-k(1+r_L)}dk. \tag{15.6}$$

Using the substitutions $s = r_L - 1$ and $s = r_L + 1$, we can express these integrals as Laplace transforms of the quantities $A^{out}(k)(1-k)$ and $A^{in}(k)(1+k)$ as follows:

$$\int_0^{\infty} A^{out}(k)(1-k)e^{-ks}dk = f(1, s+1), \tag{15.7}$$

and

$$\int_0^{\infty} A^{in}(k)(1+k)e^{-ks}dk = f(1, s-1). \tag{15.8}$$

Now, the outgoing and ingoing waves can be obtained by inverting the transforms $f(1, s+1)$ and $f(1, s-1)$, respectively, and then using the resulting expressions to evaluate the integrals in Eqs. (15.4) and (15.5).

15.1 The Sudden Turn-On: A Non-monotonic Increase in Surface Field

As our first example, we consider the expression for $f(1, r_L)$ given by

$$f(1, r_L) = \begin{cases} 0, & r_L < 1 \\ (r_L - 1)\left\{6 - (r_L - 1)^2\right\}/4, & 1 \le r_L \le 3 \\ 1, & r_L > 3, \end{cases} \tag{15.9}$$

obtained by setting $r = 1$ in our original equation for $f(r, r_L)$ (Eq. (2.10)). Substituting $s = r_L - 1$ and $s = r_L + 1$ in Eq. (15.9), and inverting the resulting Laplace transforms, we obtain the amplitudes:

$$(1 - k)A^{out}(k) = \frac{3}{2}\delta'(k) - \frac{1}{4}\delta'''(k), \tag{15.10a}$$

$$(1 + k)A^{in}(k) = -\delta(k) - \frac{3}{2}\delta'(k) + \frac{3}{2}\delta''(k) - \frac{1}{4}\delta'''(k). \tag{15.10b}$$

These Dirac delta functions have the useful property that $\int_0^\infty g(k)\delta^n(k)dk = (-1)^n g^n(0)$ for any function $g(k)$. Here, n refers to the number of derivatives that are involved in each delta function and in the general function g. This guarantees that Eqs. (15.10a) and (15.10b) can be used to evaluate the integrals in Eqs. (15.4) and (15.5). Doing this, we recover the original f-values from Eq. (2.10):

$$f(r, r_L) = \left\{3(r^2 + 1)(r_L - 1) - (r_L - 1)^3 \pm 2(r^3 - 1)\right\}/4, \tag{15.11}$$

with the minus and plus signs referring to the outgoing and ingoing waves, respectively. Thus, if we know how B_r (or A_ϕ) vary with time on the surface of the sphere, then we can determine how they vary with time at all radial distances. This was relatively easy to do in this example because we already knew the answer and it provided the $f(1, r_L)$ value given in Eq. (15.9). But what if we did not know the answer? Somehow, we would have to find a way of determining $f(1, r_L)$.

15.2 A Limited Monotonic Increase in the Strength of the Surface Field

As a second example, let us consider the expression for $f(1, r_L)$ given by

$$f(1, r_L) = \begin{cases} 0, & r_L < 1 \\ 3\{(r_L - 1)/2\}^2 - 2\{(r_L - 1)/2\}^3, & 1 \le r_L \le 3 \\ 1, & r_L > 3. \end{cases} \tag{15.12}$$

Here, $f(1, r_L)$ and the corresponding radial field strength (in units of B_0) increases gradually from 0 to 1 as r_L increases from 1 to 3, and then remains 1 afterwards. This is the symmetric cubic variation that is shown in the left panels of Figs. 7.11 and 10.1.

Proceeding as before, we substitute the relations $s = r_L - 1$ and $s = r_L + 1$ into Eq. (15.12) and invert the resulting Laplace transforms to obtain:

$$(1-k)A^{out}(k) = \frac{3}{4}\delta''(k) - \frac{1}{4}\delta'''(k), \tag{15.13a}$$

$$(1+k)A^{in}(k) = 5\delta(k) - 6\delta'(k) + \frac{9}{4}\delta''(k) - \frac{1}{4}\delta'''(k). \tag{15.13b}$$

Then using these expressions to evaluate the integrals in Eqs. (15.4) and (15.5), we obtain

$$f^{out}(r, r_L) = 1 + (1/4)(2 - 2r - r_L)(-2 - r + r_L)^2, \tag{15.14a}$$

$$f^{in}(r, r_L) = (1/4)(2 + 2r - r_L)(-2 + r + r_L)^2. \tag{15.14b}$$

Whereas the ingoing wave vanishes at its leading edge, $r = 2 - r_L$, the outgoing wave does not vanish at its leading edge, $r = r_L$. Instead, $f^{out}(r_L, r_L) = 3(1 - r_L)$. Thus, to satisfy this boundary condition, we must add another term of the form $c_1(1-r)\exp(r - r_L)$ to $f^{out}(r, r_L)$, where c_1 is a constant to be determined. Setting the combined expression for $f^{out}(r_L, r_L)$ equal to zero, we find that $c_1 = -3$. With this correction, Eqs. (15.14a) and (15.14b) are

$$f^{out}(r, r_L) = 1 + (1/4)(2 - 2r - r_L)(-2 - r + r_L)^2 - 3(1 - r)e^{r - r_L}, \tag{15.15a}$$

$$f^{in}(r, r_L) = (1/4)(2 + 2r - r_L)(-2 + r + r_L)^2. \tag{15.15b}$$

Also, when the ingoing wave arrives at the center of the sphere, another wave starts outward, adding its contributions to the fields left behind the other two waves. Thus, behind this second wave, the net f-values are

$$f^{out}(r, r_L) - f^{in}(-r, r_L) = 1 - 3(1 - r)e^{r - r_L}, \tag{15.16a}$$

$$f^{in}(r, r_L) - f^{in}(-r, r_L) = r^3, \tag{15.16b}$$

outside and inside the sphere, respectively. Thus, after $r_L = 3$, the field inside the sphere is uniform and the f-value on its surface is 1, independent of time. This is the same behavior that we obtained for the field of the exponential current increase when the rate of increase, α, is set equal to 1. In that case, the f-values in Eqs. (15.15) and (15.16) are the same as the ones in Eqs. (7.14) and (7.15).

Also, these fields have the same source currents, as we can show by evaluating the current associated with the f-values in Eqs. (15.15) and (15.16). As discussed previously, the azimuthal current per unit distance in the meridional direction is given by the jump of B_θ/μ_0 at $r = 1$. Recalling that $B_\theta = (B_0/2)r^{-3}(f - rf')\sin\theta$ and that f is continuous across $r = 1$, the jump of B_θ/μ_0 becomes $-(B_0/\mu_0)(1/2)(f_o' - f_i')\sin\theta$, where f_o and f_i are the f-values outside and inside the sphere. Using the f-values given by Eqs. (15.15a) and (15.15b), we obtain the current per unit length

$$\int J_\phi dr = \frac{3}{2}\frac{B_0}{\mu_0}\left\{1 - e^{-(r_L-1)}\right\}\sin\theta, \tag{15.17}$$

valid for r_L in the range $(1, 3)$. The same result is obtained from Eqs. (15.16a) and (15.16b) when $r_L > 3$. This current profile is identical to that shown in the left panel of Fig. 7.6 when $\alpha = 1$ and $\xi = r_L - 1$. Thus, the symmetric cubic variation of $f(1, r_L)$ given by Eq. (15.12) corresponds to a current ramp that rises exponentially with an initial slope of unity. By comparison, if we had done the same calculation using the f-values in Eq. (15.11) (and its extension to $f_o = 1$ and $f_i = r^3$ behind the bubble), we would have obtained

$$\int J_\phi dr = \begin{cases} 0, & r_L < 1 \\ (3/2)(B_0/\mu_0)\sin\theta, & r_L \geq 1, \end{cases} \tag{15.18}$$

corresponding to the sudden turn-on of current given in Eq. (2.1).

Figure 15.1 shows the f-values and currents for these two examples (solid red and blue curves). For comparison, I have included similar results for exponential current increases at the rates $\alpha = 2.5$ (dashed) and $\alpha = 0.5$ (dotted), which correspond to rise times of $0.4R/c$ and $2.0R/c$, respectively. The latter profiles were obtained by plotting the f-values given in Eq. (7.15) with $r = 1$. Note that as the rise time decreases (and the rate $\alpha \to \infty$), the exponential current source approaches the step function of the sudden turn-on, and the f-value for that exponential current source approaches the same peak at $r = 1 + \sqrt{2} \approx 2.41$.

Referring to Eq. (15.16a), it is interesting that the field B_r at the surface of the sphere reaches its final value at the time, $r_L = 3$, that the second wave breaks through the surface, and does not change afterward despite the continued increase of current toward its asymptotic value. Also, at a given location outside the sphere, the field continues to change asymptotically with time after the field on the surface has reached its final value and sufficient time has passed for this information to

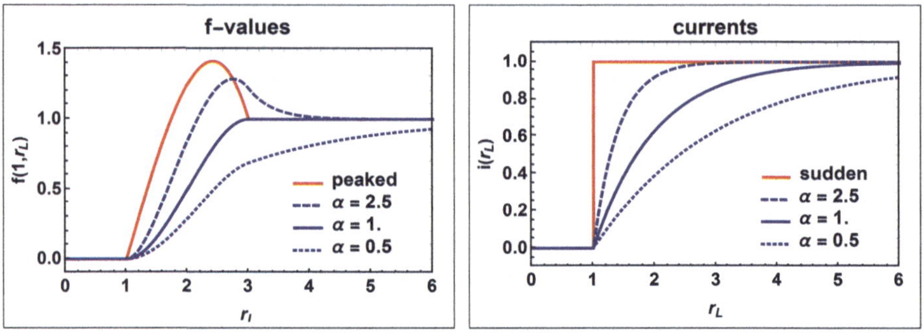

Fig. 15.1 Time profiles of f-values and electric currents (normalized to $(3/2)(B_0/\mu_0)$) on the surface of a sphere when the current is turned on suddenly (solid red) and exponentially with three different rise times. The solid red and blue curves correspond to the cubic f-values given by Eqs. (15.9) and (15.12), respectively. $f(1, r_L)$ is the magnetic flux (in units of Φ_0) crossing the sphere as a function of time

propagate to the field point. Thus, although the bubble has a sharp leading edge, it does not have a sharp trailing edge.

15.3 An Unlimited Monotonic Increase of Surface Field

Finally, as a third example of a boundary-value problem, we consider the case of a surface flux density, $B_r(1, t)$, that grows linearly with time. For a dipole field, the boundary condition on B_r is

$$B_r(1, t) = B_0\, \alpha t \cos\theta, \tag{15.19}$$

where α is the growth rate of the field. Intuitively, we expect the magnetic field strength to be linearly proportional to t, and the electric field, whose curl is the negative time derivative of the magnetic field, to be static.

We begin by transforming the boundary condition on B_r to a condition on A_ϕ. This is easily done for our axisymmetric field whose radial component is simply

$$B_r = \frac{1}{r \sin\theta} \frac{\partial(A_\phi \sin\theta)}{\partial\theta}. \tag{15.20}$$

With this equation, we can see that the condition on B_r given in Eq. (15.19) reduces to the following condition on A_ϕ:

$$A_\phi(1, t) = \left(\frac{B_0 R}{2}\right) \alpha t \sin\theta. \tag{15.21}$$

As before, the $\sin\theta$-dependence means that $\ell = 1$.

If we assume that $A_\phi = f(r, r_L)\sin\theta/r^2$, then $f(r, r_L)$ must satisfy

$$f'' - \frac{2}{r}f' - \frac{\partial^2 f}{\partial r_L^2} = 0. \quad (15.22)$$

From this equation, it is easy to see that if $f(r, r_L)$ depends linearly on r_L, as $f(r, r_L) \sim \alpha r_L g(r)$ for example, then the remaining spatial factor, $g(r)$, satisfies the even simpler equation

$$g'' - \frac{2}{r}g' = 0, \quad (15.23)$$

whose solutions are $g(r) = r^3$ and $g(r) = 1$, valid inside and outside the sphere, respectively. The corresponding vector potential and field components are

$$A_\phi(r, \theta, t) = \left(\frac{B_0 R}{2}\right)(\alpha t)\left(\frac{r}{R}\right)\sin\theta, \quad (15.24a)$$

$$B_r(r, \theta, t) = B_0\,(\alpha t)\cos\theta, \quad (15.24b)$$

$$B_\theta(r, \theta, t) = -B_0\,(\alpha t)\sin\theta, \quad (15.24c)$$

$$E_\phi(r, \theta) = -\frac{1}{2}\left(\frac{\alpha R}{c}\right)(cB_0)\left(\frac{r}{R}\right)\sin\theta, \quad (15.24d)$$

inside the sphere, and

$$A_\phi(r, \theta, t) = \left(\frac{B_0 R}{2}\right)(\alpha t)\left(\frac{r}{R}\right)^{-2}\sin\theta, \quad (15.25a)$$

$$B_r(r, \theta, t) = B_0\,(\alpha t)\left(\frac{r}{R}\right)^{-3}\cos\theta, \quad (15.25b)$$

$$B_\theta(r, \theta, t) = \frac{B_0}{2}\,(\alpha t)\left(\frac{r}{R}\right)^{-3}\sin\theta, \quad (15.25c)$$

$$E_\phi(r, \theta) = -\frac{1}{2}\left(\frac{\alpha R}{c}\right)(cB_0)\left(\frac{r}{R}\right)^{-2}\sin\theta, \quad (15.25d)$$

outside the sphere.

These equations are similar to those of the static dipole field given by Eqs. (3.4) and (3.5). The main differences are that the vector potential and the magnetic fields are multiplied by the factor αt, which gives them a linearly increasing time dependence, and that the electric fields are replaced by static terms given by $E_\phi = -\partial A_\phi/\partial t$. A more precise comparison is with the fields of the middle region for the linear current ramp in Chap. 5. The equations are the same when the slope of the current ramp $(1/\Delta r)$ is replaced by the normalized growth rate, $\alpha R/c$, of the field on the surface of the sphere, and the starting time of the ramp is shifted from $r_L = 1$ to $r_L = 0$.

Despite the presence of an electric field, there is no surface charge on the sphere because $E_r = 0$ everywhere. Also, there is no volume charge because $\nabla \cdot \mathbf{E} = \partial E_\phi / \partial \phi = 0$. However, the jump of B_θ across the sphere gives a surface current whose value is αt times the value that we already found for the static dipole field. Similarly, the magnetic fields in Eqs. (15.24) and (15.25) are αt times as strong as those of the uniformly magnetized sphere. The electric fields are static with a strength that is reduced from cB_0 by the factor $\alpha R/c$, which is small unless the magnetic field changes appreciably in a time R/c.

Finally, like the fields for the linear current ramp, these fields have a Poynting flux, $S_r = E_\phi B_\theta / \mu_0 \sim r^{-5}$, which gives no net radiation across a spherical surface far from the source. The flux simply circulates meridionally in the $r\theta$ planes, but does not escape. Only "accelerated" charges, like those obtained by turning the current on or off, produce the bubbles which carry energy away from the sphere.

Chapter 16
The Conducting Core

Here, we consider what happens when the current in a unit sphere is turned on suddenly while a smaller perfectly conducting sphere of radius ϵ is centered inside. As in the case without the central core, two waves will leave the current source, one heading outward away from the sphere and the other heading inward toward the center. If the core were not present, the incoming wave would turn around when it reached the center and then move back out through the sphere to form the trailing edge of an outgoing bubble of thickness $2R$ (where $R = 1$ for the unit sphere).

The infinite conductivity of the central core will prevent the magnetic field from penetrating it and cause the ingoing wave to turn around prematurely. Thus, the separation between the outgoing waves would be smaller by the amount, 2ϵ, and a static, dipole-like field would be left behind, rooted, not in the unit sphere, but in the portion of the unit sphere external to the conducting core. Intuitively, this remnant field would be equivalent to the static field of two uniformly magnetized spheres having the same central field strength, but opposite polarities, similar to the field that is shown in the left panel of Fig. 14.4. Now, let us see what a calculation actually gives.

Three waves will contribute to the field. Outside the unit sphere, there will be an outgoing wave associated with the leading edge of the disturbance: $f_L^{out}(r, r_L) = f_o(r, r_L)$, where f_o is the external f-value given by Eq. (2.11a). Inside the unit sphere, there will be an ingoing wave associated with the following edge of the disturbance: $f_F^{in}(r, r_L) = f_i(r, r_L)$, where $f_i(r, r_L)$ is the internal f-value given by Eq. (2.11b). Also, inside the unit sphere, there will be an outgoing wave, $f_F^{out}(r, r_L)$, produced by the reflection of the ingoing wave at the conducting core. Our first objective will be to determine this outgoing wave.

Electronic Supplementary Material The online version of this chapter (https://doi.org/10.1007/978-3-030-40264-8_16) contains supplementary material, which is available to authorized users.

N. R. Sheeley, Jr., *Transient Magnetic Fields*,
https://doi.org/10.1007/978-3-030-40264-8_16

The outgoing wave, $f_F^{out}(r, r_L)$, must satisfy three conditions. First, it must be constructed from outgoing solutions to the wave equation (15.22). As we have seen in Eq. (15.4), these solutions are generally of the form $(1 - kr/\epsilon)\exp\{k(r - r_L)/\epsilon\}$, where k is a constant. Second, the outgoing wave must combine with the ingoing wave to make the radial component of magnetic field vanish at the surface of the core, so that $B_r(\epsilon, r_L) = 0$ for all r_L. This is equivalent to

$$f_F^{in}(\epsilon, r_L) + f_F^{out}(\epsilon, r_L) = f_i(\epsilon, r_L) + f_F^{out}(\epsilon, r_L) = 0. \tag{16.1}$$

Third, the outgoing wave must have a "leading edge" at $r = r_L - 2(1 - \epsilon)$, given by

$$f_F^{out}(r_L + 2\epsilon - 2, r_L) = 0, \tag{16.2}$$

for all $r_L \geq 2 - \epsilon$.

We can use these conditions to determine $f_F^{out}(r, r_L)$ as follows. We begin with the particular combination of outgoing waves:

$$f_F^{out}(r, r_L) = \int_0^\infty A(k)dk(1 - kr/\epsilon)e^{k(r - r_L - 2\epsilon + 2)/\epsilon} + c_1(1 - r/\epsilon)e^{(r - r_L - 2\epsilon + 2)/\epsilon}, \tag{16.3}$$

where c_1 is a constant to be determined. In this equation, the continuum spectrum, $A(k)$, allows us to satisfy the boundary condition (16.1), and the "line" at $k = 1$ allows us to satisfy the leading-edge condition (16.2). Thus, when we put $r = \epsilon$ in Eq. (16.3), the c_1-dependent term vanishes and the resulting expression for $f_F^{out}(\epsilon, r_L)$ may be equated to $-f_i(\epsilon, r_L)$, according to Eq. (16.1). The result is

$$\int_0^\infty A(k)(1 - k)e^{-sk}dk = -f_i\{\epsilon, (s - 1)\epsilon + 2\}, \tag{16.4}$$

where $s = (r_L + \epsilon - 2)/\epsilon$. Recognizing that the left term is the Laplace transform of $A(k)(1 - k)$ with respect to the variable s, and inverting this transform, we obtain $A(k)(1 - k)$ as a series of delta functions and their derivatives:

$$A(k)(1 - k) = -\frac{3}{2}\epsilon^2\,\delta'(k) + \frac{3}{4}\epsilon^2(1 - \epsilon)\,\delta''(k) + \frac{1}{4}\epsilon^3\,\delta'''(k). \tag{16.5}$$

Equation (16.2) tells us that if we set $r = r_L + 2\epsilon - 2$ in Eq. (16.3), the result must vanish for all $r_L \geq 2 - \epsilon$. This can be true only if the constant c_1 is given by

$$c_1 = -\int_0^\infty A(k)kdk = -\int_0^\infty A(k)dk = -3\epsilon^2(1 - \epsilon) = \epsilon^3 + f_i\{0, 2(1 - \epsilon)\}. \tag{16.6}$$

Substituting these values of $A(k)$ and c_1 into Eq. (16.3), we finally obtain the expression for $f_F^{out}(r, r_L)$:

$$f_F^{out}(r, r_L) = r^3 - \epsilon^3 - f_i(r, r_L) + 3\epsilon(1-\epsilon)(r-\epsilon)e^{-\{r_L - 2(1-\epsilon) - r\}/\epsilon}. \quad (16.7)$$

After the following edge has passed out through the unit sphere and the bubble has moved outward, the residual f-values inside and outside the unit sphere are given by $f_i(r, r_L) + f_F^{out}(r, r_L)$ and $f_o(r, r_L) + f_F^{out}(r, r_L)$, respectively. We can express these values as

$$f(r, r_L) = \begin{cases} r^3 - \epsilon^3 + 3\epsilon(1-\epsilon)(r-\epsilon)e^{-\{r_L - 2(1-\epsilon) - r\}/\epsilon}, & \epsilon \le r \le 1 \\ 1 - \epsilon^3 + 3\epsilon(1-\epsilon)(r-\epsilon)e^{-\{r_L - 2(1-\epsilon) - r\}/\epsilon}, & 1 \le r \le r_L - 2(1-\epsilon). \end{cases} \quad (16.8)$$

The f-value is still $f_o(r, r_L)$ between the two outgoing waves, but now this region extends only $2(1-\epsilon)$ units behind the leading edge compared to 2 units behind the leading edge of the bubble produced in the absence of the conducting core. The difference, 2ϵ, corresponds to the length of the path that the ingoing wave did not take. Note that the first part of Eq. (16.8) is valid behind the trailing edge as soon as that wave reflects off the conducting core and heads outward, but the second part of the equation does not apply until the trailing edge crosses the surface at $r = 1$.

Referring to Eq. (16.8), we see that the fields left behind the bubble are not static when a conducting core is present inside the unit sphere. However, the time-varying terms decay exponentially and become greatly reduced when the field points are more than a few units of ϵ behind the following edge of the bubble. In that case, $f(r, r_L) \to r^3 - \epsilon^3$ inside the unit sphere, and $f(r, r_L) \to 1 - \epsilon^3$ outside. As we guessed initially, these f-values represent the field produced by two uniformly magnetized spheres of equal strength, but opposite polarity, one of radius ϵ, and the other of radius 1. This field is identical to the static field whose field lines are plotted in the left panel of Fig. 14.4. In both cases, the field vanishes inside the inner sphere and is a weakened dipole outside the outer sphere. In the transition region between the two spheres, the small dipole contribution causes the uniform field lines to bend smoothly around the inner sphere.

Figure 16.1 compares the evolution of the field lines for a sudden turn-on without a conducting core (top panels) and the evolution with conducting cores of radii $\epsilon =$ 1/3, 1/2, and 2/3 shown in the second, third, and fourth rows, respectively. The dotted circles mark the positions of the leading and trailing waves at the times $r_L = 2.6$, 3.6, and 4.6. In each case, the spatial scale is varied to follow the leading edge of the outgoing wave. For progressively larger conducting cores of radius, ϵ, the separation between the leading and trailing waves shows the expected decrease as $2(1-\epsilon)$. At $r_L = 2.6$ and $\epsilon = 0$, the ingoing wave has turned around and is moving outward through the sphere at the location $r = 0.6$. For $\epsilon \ge 1/3$, the wave is already outside the sphere trailing behind the leading wave. As expected for $\epsilon > 0$, the field lines inside the sphere stream smoothly around the conducting core and have no components normal to its surface.

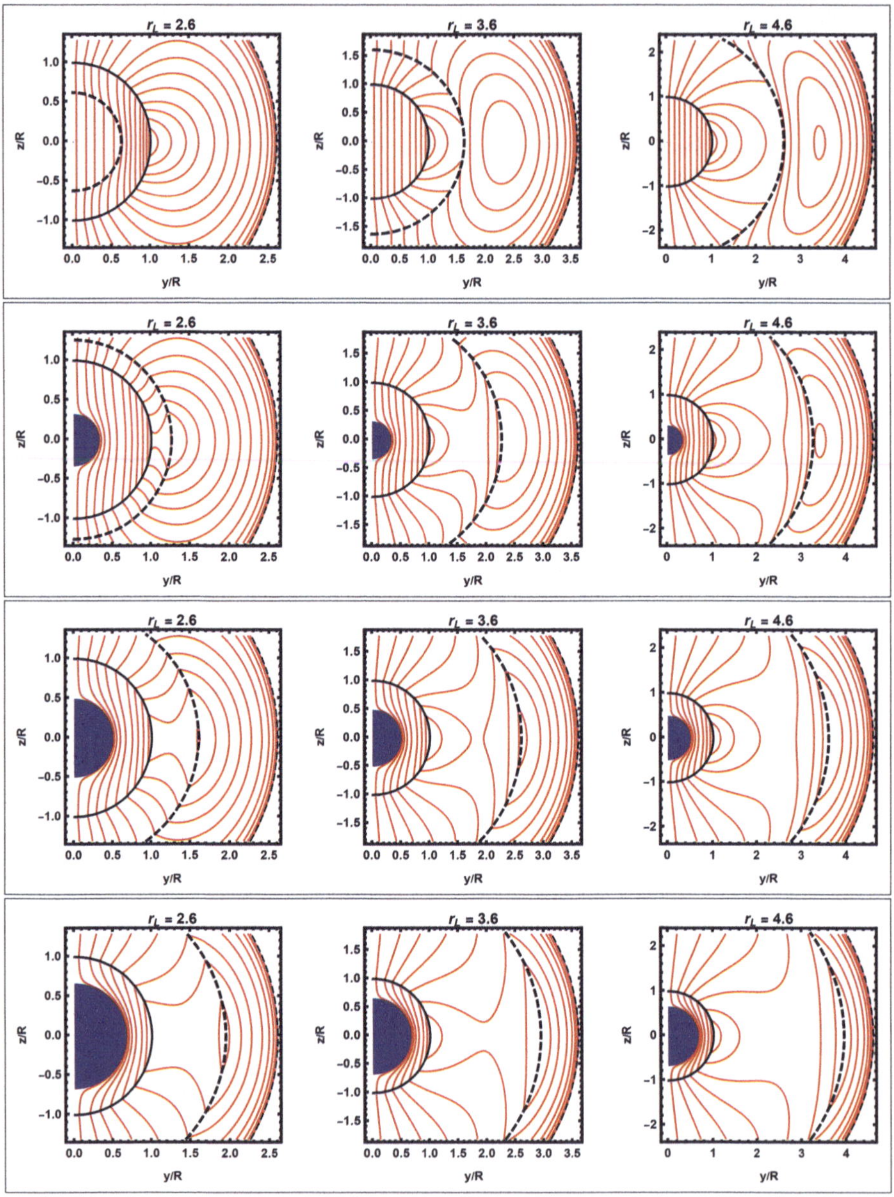

Fig. 16.1 Field lines for sudden turn-ons with conducting cores of radius $\epsilon = 0, 1/3, 1/2, 2/3$ (top to bottom). The wave fronts and conducting cores are indicated by dashed circles and blue disks, respectively

These maps reveal two additional features of the field when a conducting core is present. First, an X-type neutral point (line in three dimensions) forms well behind the reflected wave front, and marks the end of the region of detached flux (i.e., the back end of the bubble). As a result, this region of detached flux falls behind the two waves and becomes increasingly larger with time. Second, between the two wave fronts, the O-type neutral point, familiar to us when $\epsilon = 0$, lies progressively closer to the following boundary at $r = r_L - 2(1-\epsilon)$ as ϵ increases from 0, and the circulating contours become significantly flattened for values of $\epsilon > 1/2$.

For the case of $\epsilon = 0$, we have already seen in Chap. 3 that the O-type neutral point moves toward the center of the bubble and asymptotically approaches the location $r = r_L - 1$. This asymptotic location will fall behind the trailing wave at $r = r_L - 2(1-\epsilon)$ when $\epsilon \geq 1/2$. In this case, we might expect the neutral point to disappear, changing to a discontinuity analogous to the one that lies at the trailing edge of the bubble when $\epsilon = 0$.

It is interesting to examine these discontinuities in detail. We begin with $\epsilon = 0$ and $r = 1$. Dropping the common factor of $\sin\theta$, we find that B_θ jumps from -1 (corresponding to the uniform field, B_0, in the sphere) to $+1/2$ for a total discontinuity of $+3/2$ (corresponding to the strength of the source current on the sphere). Similar jumps of $+3/2$ are also present at $r = 1$ when $\epsilon > 0$. The jump at $r = r_L - 2(1-\epsilon)$ is $\Delta B_\theta / B_0 = +3/(4r)$ (corresponding to $\Delta E_\phi/(cB_0) = -3/(4r)$), and the jump at $r = r_L$ is $\Delta B_\theta / B_0 = -3/(4r)$ (corresponding to $\Delta E_\phi/(cB_0) = +3/(4r)$). As mentioned in Chap. 3, in the absence of currents, the jumps of B_θ and E_ϕ are related by $\Delta E_\phi = -c\Delta B_\theta$.

In the presence of a conducting core, there is also a discontinuity at the surface of the core where $r = \epsilon$. Using the f-value given in Eq. (16.8) with $\epsilon \leq r \leq 1$, it is easy to show that when $r = \epsilon$, the jump is

$$\frac{\Delta B_\theta}{B_0} = \frac{1}{2} r^{-3} \Delta(f - rf') = -\frac{1}{2} r^{-2} \Delta(f')$$
$$= -\frac{3}{2}\left\{1 + \left(\frac{1-\epsilon}{\epsilon}\right) e^{-\{r_L - (2-\epsilon)\}/\epsilon}\right\}, \quad (16.9)$$

again dropping the common factor of $\sin\theta$. Here, we see that the jump of B_θ / B_0 approaches $-3/2$ as $r_L \to \infty$. If we restore the $\sin\theta$-dependence of B_θ and then integrate B_θ over the surface of the conducting core of radius ϵ, then we obtain the total induced current on that surface as a function of time:

$$i(r_L) = \begin{cases} 0, & r_L \leq 2(1-\epsilon), \\ -3(B_0 R/\mu_0)\left[\epsilon + (1-\epsilon)\, e^{-\{r_L - (2-\epsilon)\}/\epsilon}\right], & r_L > 2(1-\epsilon). \end{cases} \quad (16.10)$$

From this equation, we see that the induced current begins with the value $-3B_0 R/\mu_0$ and approaches the asymptotic value of $-(3B_0 R/\mu_0)\epsilon$ on an exponential time scale of $\epsilon R/c$. Thus, after a few time scales, the induced current is similar to the current on the outer sphere, except that the induced current is of opposite sign and is smaller because it scales as the size of the conducting core. Also, note that as $\epsilon \to 1$, the induced current loses its time dependence, and maintains a static value of $-(3B_0R/\mu_0)\epsilon$.

When they exist, the neutral points are determined by setting $B_\theta(r, \pi/2, r_L) = 0$, which is equivalent to $f - rf' = 0$. The O-type neutral point is obtained by substituting $f = f_o$, which gives the result that we obtained in Chap. 3 in the absence of the conducting core (i.e., $r \to r_L - 1$ as $r_L \to \infty$). However, as we mentioned above, this solution is valid only when $\epsilon < 1/2$. There is no O-type neutral point when $\epsilon \geq 1/2$, and the detached field lines circulate around the discontinuity at $r = r_L - 2(1 - \epsilon)$.

The X-type neutral point is obtained by substituting f from Eq. (16.8) with $1 \leq r \leq r_L - 2(1 - \epsilon)$, which gives

$$f - rf' = 1 - \epsilon^3 - 3(1 - \epsilon)\left(\frac{r^3 + \epsilon^3}{r + \epsilon}\right)e^{-\{r_L - 2(1-\epsilon) - r\}/\epsilon} = 0, \tag{16.11}$$

which can be solved for r_L as a function of r:

$$r_L = 2(1 - \epsilon) + r + \epsilon\, ln\left\{3\left(\frac{1 - \epsilon}{1 - \epsilon^3}\right)\left(\frac{r^3 + \epsilon^3}{r + \epsilon}\right)\right\} \tag{16.12}$$

For $\epsilon \ll 1$ and $r \gg 1$, this equation has the approximate solution:

$$r_L - 2(1 - \epsilon) - r \approx \epsilon\, ln(3r^2), \tag{16.13}$$

which means that

$$r \approx r_L - 2(1 - \epsilon) - \epsilon\, ln(3r^2), \tag{16.14}$$

and the neutral point lies a distance $\epsilon\, ln(3r^2)$ behind the trailing wave. Thus, as time increases, the neutral point falls increasingly farther behind the trailing wave, while still moving outward from the sphere, and the thickness of the bubble increases accordingly.

Next, we consider the flux budget for this sudden turn-on. The amount of flux that is ultimately contained in the external dipole field is $\Phi_e/\Phi_0 = f_o(r_L, r_L)/r_L - f(1, r_L)$, where $\Phi_0 = \pi R^2 B_0$ with $R = 1$, and $f(1, r_L)$ is given by Eq. (16.8). The first term vanishes because it refers to f_o at the leading edge of the disturbance. The second term is obtained by substituting the expression for r_L given in Eq. (16.12) into Eq. (16.8) and letting $r_L \to \infty$. The result is $\Phi_e/\Phi_0 = -(1 - \epsilon^3)$, where the minus sign indicates that B_θ is positive. Equivalently, the corresponding amount of flux trapped in the sphere is $\Phi_t/\Phi_0 = 1 - \epsilon^3$.

Next, we consider the radiated flux. If $\epsilon \leq 1/2$, then the amount of detached flux is $\Phi_d/\Phi_0 = f_o(r_L - 1, r_L)/(r_L - 1) - f(r, r_L)/r$, where $f(r, r_L)$ is given by Eq. (16.8) with $1 \leq r \leq r_L - 2(1 - \epsilon)$, and r is related to r_L by Eq. (16.12). As we found in Chap. 3, the first term approaches 3/4 as $r_L \to \infty$. The second term vanishes when $r \to \infty$. Consequently, for $\epsilon \leq 1/2$, the amount of radiated flux is $\Phi_r/\Phi_0 = 3/4$, just as we found in the absence of a conducting core.

However, if $\epsilon > 1/2$, there is no O-type neutral point and the flux circulates around the location $r = r_L - 2(1-\epsilon)$. In this case, the amount of detached flux is $\Phi_d/\Phi_0 = f[r_L - 2(1-\epsilon), r_L]/[r_L - 2(1-\epsilon)] - f(r, r_L)/r$. As before, the second term vanishes and the first term approaches $3\epsilon(1-\epsilon)$. Thus, the radiated flux $\Phi_r/\Phi_0 = 3\epsilon(1-\epsilon)$ for $\epsilon > 1/2$, which matches the value of 3/4 at $\epsilon = 1/2$, and falls continuously to 0 as ϵ increases to 1. With a little algebra, we can summarize this result as:

$$\frac{\Phi_r}{\Phi_0} = \begin{cases} 3/4, & 0 \leq \epsilon \leq 1/2 \\ (3/4) - 3(\epsilon - 1/2)^2, & 1/2 < \epsilon \leq 1. \end{cases} \tag{16.15}$$

The conducting core does not affect the radiated flux unless its radius, ϵ, exceeds 1/2, in which case the amount of radiated flux is smaller than it is without the core.

Next, we consider the energy budget. To evaluate the total energy provided by the current source, we return to the volume integral given by Eq. (4.6), but with the limits of r_L-integration changed from (1, 3) to (r_{L1}, r_{L2}). This integral gives the total energy provided by the source between the times r_{L1} and r_{L2}:

$$\frac{\mathcal{E}_{tot}}{\mathcal{E}_0} = 3\{f(1, r_{L2}) - f(1, r_{L1})\}, \tag{16.16}$$

where $\mathcal{E}_0 = (4\pi R^3/3)(B_0^2/2\mu_0)$ is the field energy in a uniformly magnetized sphere of radius, R. For our source at $r = 1$ with the conducting core at $r = \epsilon$, we set $r_{L1} = 1$ and recognize that $f(1, 1) = f_o(1, 1) = 0$ so that the second term in Eq. (16.16) vanishes. Then, we let $r_{L2} \to \infty$ and evaluate $f(1, r_{L2})$ from Eq. (16.8) to obtain

$$\mathcal{E}_{tot} = 3(1-\epsilon^3)\mathcal{E}_0 = 3\left\{\frac{4}{3}\pi R^3 - \frac{4}{3}\pi(\epsilon R)^3\right\}\left(\frac{B_0^2}{2\mu_0}\right), \tag{16.17}$$

where $R = 1$ for the unit sphere. Thus, the total energy provided by the source is reduced by three times the energy that would have been in the volume now occupied by the conducting core. This reduction is equal to the energy difference, $\Delta\mathcal{E} = M_i B_0 = 3\mathcal{E}_i$, obtained from Eq. (14.25) when $B_i = -B_o$, for the field of anti-parallel dipole moments shown in the left panel of Fig. 14.4.

But how much of this energy is stored in the field and how much is radiated away? Returning this time to Eq. (4.3), we set $r_{L1} = r$ and let $r_{L2} \to \infty$ to obtain the Poynting energy that ultimately crosses the surface at radius r. For the f-value given by Eq. (16.8) with $1 \leq r \leq r_L - 2(1-\epsilon)$, we obtain

$$\frac{\mathcal{E}}{\mathcal{E}_0} = \left\{\frac{3}{2} + \frac{1}{2r^3}\right\}(1 - \epsilon^3), \tag{16.18}$$

which approaches $\mathcal{E}_{rad}/\mathcal{E}_0 = (3/2)(1 - \epsilon^3)$ as $r \to \infty$. Thus, like the case without a conducting core, half the energy is radiated and therefore half is stored in the static magnetic field that is ultimately left behind. A separate evaluation of the field energy gives $(3/2)(1 - \epsilon^3)$, of which $(1 - \epsilon^3)\{3/2 - (1/2)(1 - \epsilon^3)\}$ lies inside the sphere of unit radius, and $(1/2)(1 - \epsilon^3)^2$ lies outside. For $\epsilon \ll 1$, this corresponds to $1 - (1/2)\epsilon^3$ inside and $(1/2) - \epsilon^3$ outside. Thus, of the total field energy deficit, $(3/2)\epsilon^3$, caused by the conducting core, approximately 2/3 is removed from outside the sphere and 1/3 from inside the sphere. Of course, no field energy lies inside the conducting core.

All of these energies approach 0 as $\epsilon \to 1$, indicating that little energy is transferred to either the field or the radiation when only a small space is left between the current source and the central conducting core. Also, unlike the radiated flux, the radiated energy is diminished by a finite amount even for cores with $\epsilon < 1/2$. However, this is a small effect, varying as ϵ^3, and when $\epsilon = 0$, we regain all of the energies previously obtained without the core: 1 unit stored inside the sphere, 1/2 stored outside, and 3/2 radiated away. Of course, the radiated energy would be smaller if the current were turned on gradually.

Chapter 17
Summary

I began this book with an idealized model in which the electric current on a sphere was turned on suddenly, and then I used Maxwell's equations to derive the properties of the resulting magnetic and electric fields. Thus, to the extent that the derivations are error-free, we can be sure that these fields and their properties are correct. They are all direct consequences of Maxwell's equations. However, the book contains many results, and they are described in the precise, mathematical language that was used to find them.

In this chapter, I will try to summarize these results in physical terms without using the mathematical description (or at least without using very much of it). A successful physical interpretation would solidify our understanding and place the diverse results in an overall context that we can remember. However, such an interpretation will involve guesses and generalizations that may turn out to be wrong. That is the danger of not using a precise mathematical description. But, if we are lucky, most of the guesses and generalizations will be correct, and at the same time they will provide the insight and improved understanding that we desire.

17.1 The Model

Let us begin with the model. We started by suddenly turning on an electric current. The current was "wrapped around" the surface of a sphere with a $\sin\theta$ distribution, where θ is co-latitude. This distribution produces a uniform field inside the sphere and a dipole field outside the sphere. By turning this current on suddenly, we produced two waves that moved away from the current. One wave moved outward from the sphere. The other wave moved inward toward the center where it turned around and headed outward again. Eventually, it passed through the surface of the sphere and followed the original outgoing wave into the distance. (An equivalent interpretation was for the inward wave to pass through the center and move outward

N. R. Sheeley, Jr., *Transient Magnetic Fields*,
https://doi.org/10.1007/978-3-030-40264-8_17

on the opposite side of the sphere.) As a result, there were two outgoing waves, one with its meridional field directed downward toward the equator and another with its meridional field directed upward. Together these upward and downward fields formed a bubble of concentric loops carrying magnetic flux and energy away from the sphere.

In retrospect, I suppose that we would have produced essentially the same result by letting all of the infinitesimal loops on the sphere slip down to the equator to form a single equatorial loop, and then suddenly turning on the current in that loop. A disturbance would expand from the current loop, producing both ingoing and outgoing components. The ingoing disturbance would turn around at the center (or pass through the ingoing component from the opposite side of the current loop) and move outward after the original outgoing disturbance. The oppositely directed meridional fields of these two waves would form a bubble of circulating flux. At large distances from the current loop, the fields would be indistinguishable from those of the spherical current source, but the amounts of radiated flux and energy would be different because they depend on the entire field and not just its distant component.

Thus, it seems plausible that any finite current distribution is equivalent to a current loop, and that by turning this current on suddenly, we would generate a toroidal bubble of concentric loops, carrying magnetic flux and energy away from the loop. We would not have obtained this result by analyzing the field produced by the current in a straight wire of infinite length.

17.2 Radiated Flux and Energy

In Chaps. 3 and 4, we learned that when the current on the sphere was turned on suddenly, the resulting bubble carried away 3/4 of a unit of flux and 3/2 of a unit of energy. As mentioned in the previous subsection, it seems likely that those numbers are a consequence of the detailed current distribution on the surface of the sphere, and that the sudden turn-on of the current in a flat loop would give different values of radiated flux and energy. However, regardless of how much energy is radiated from the flat loop, I imagine that an identical amount of energy would be left in the field if the current were turned on suddenly. That way, no extra energy would be required when the current is suddenly turned off; the new bubble would sweep up all the field energy and depart with the same amount of radiated energy that was carried by the first bubble.

17.3 The Linear Current Ramp of Finite Duration

In Chap. 5, the analysis was applied to the field of a spherical current that increases linearly with time for a duration that is finite, but longer than the transit time across the sphere. We regarded the current ramp as a long staircase of infinitesimal steps, and then added the contributions from the individual steps. The steps that occurred during the initial transient of duration $2R/c$ produced a strong meridional field of positive polarity (i.e., pointing downward). Likewise, the steps at the end of the current ramp produced a sheath of negative-polarity field. These two sheaths of oppositely directed field formed the leading and trailing boundaries of a growing dipole field, and they left the fully grown dipole field in their wake. These sheath fields decreased more slowly with radial distance than the dipole field within the bubble, and therefore ended up carrying all of the radiated flux, energy, and momentum.

From the viewpoint of a distant observer, the growing dipole field tells the observer that the current has been turned on and is increasing with time (or was increasing with time when the wave left the sphere). Likewise, a collapsing dipole tells the observer that the current is decreasing (or was decreasing when the bubble left its current source). When the end of this bubble arrives and the dipole field disappears, the observer knows that the current has been turned off. If the current had been turned off suddenly, the field would disappear as soon as the much narrower bubble of width, $2R$, arrived and there would be no collapsing loops.

17.4 Stretching and Pinchoff

Because a changing current produces two waves of opposite meridional field, one of these waves will have a polarity opposite that of the background field, which ensures that a neutral point will always be produced in a field whose strength decreases more rapidly with distance than the wave. The initial drift of field lines backwards toward the sphere is the first clue that a neutral point will soon form. This inward motion will be slow at low heights, but the downward motion will become faster at greater heights, until it eventually creates an "evacuated" region where the legs of the long, stretched loops pinch together and reconnect.

This reconnection separates the originally stretched loop into two parts—a short, collapsing loop still attached to the sphere and a detached loop of closed flux moving away from the sphere. The trailing edge of this outgoing loop overtakes its leading edge, causing the shrinking loop to disappear at an O-type neutral point within the bubble. Eventually, a steady state occurs in which the creation of closed loops at the X-type neutral point balances the destruction of closed loops at the O-type neutral point, and this bubble of loops carries away a constant amount of magnetic flux and energy. This is the mechanism for separating an outgoing bubble of flux from the field where it originated.

An interesting aspect of this process is its bizarre analogy to electron scattering by pair production in the field of a nucleus. This analogy arises when the locations of the magnetic loops are displayed in a space-time map with its slit oriented radially outward at the "equator" of the sphere. In this case, each space-time track looks like the path of an electron in a Feynman diagram (Feynman 1949, 1985). In both cases, a threshold is involved. If the flux corresponding to a given magnetic field line is less than a particular amount, the space-time track bends "backward in time" for a while before resuming its journey forward in time. Likewise, if the energy of the incident gamma ray exceeds the combined rest mass of the electron and the positron, the path will have a segment in which the electron moves backward in time.

In the Feynman diagram, this time-reversed segment corresponds to a positron moving forward in time, and in the space-time map, it corresponds to a reversed-polarity field line rising upward from the X-type neutral point where the reconnection occurred. The positron is annihilated when it reaches the incident electron, and the reversed-polarity field line is annihilated when it arrives at the O-type neutral point and meets its counterpart from the top of the stretched loop. In this way, pair production provides a discontinuous change in the scattering angle of the incident electron, and field line reconnection transfers flux from its "height of no-return" to a location behind the outgoing bubble and closer to the sphere. If the threshold is not reached, then the electron undergoes Compton scattering with the incident gamma ray and the loop falls continuously back toward the sphere.

It is tempting to pursue this analogy further by assigning a field to the electron (which might be a wave function in quantum mechanics), and then trying to interpret the electron-positron pair production in terms of reconnection at an X-type neutral point of that field. From our more rigorous analysis of the magnetic field, it is easy, exciting, but perhaps crazy to imagine that the positron is created by the stretching and Pinchoff of a field associated with the electron. Such speculation is beyond the scope of this book, and its pursuit will be left for the curious reader.

17.5 Non-linear Current Ramps

The next question was what happens if the current ramp is not linear, but begins or ends gradually instead of abruptly at a specific time. As an example, we considered ramps that either started or ended with a quadratic profile, called quadratic/linear (QL) or linear/quadratic (LQ) for simplicity. Again, we regarded the resulting fields as a sequence of infinitesimal contributions, but with variable spacing at the quadratic (Q) end. In this case, it was easy to see that the resulting magnetic bubble would have the familiar sheath of compressed flux at the linear (L) end, but its return flux would occur in more widely spaced contours at the quadratic (Q) end.

An interesting comparison was between the fields of the LQ off-ramp and the exponentially decaying current. Because these current profiles are similar, we might expect the exponential current profile to produce a bubble with compressed negative flux in a leading sheath of thickness $2R$, and positive return flux that is distributed over a much wider radial region. This was essentially what we found. The main exception was that for the LQ profile, the return flux was spread over a finite region with no field left behind, whereas for the exponential off-ramp, the return flux was spread all the way back to a fixed point at the top of the decaying dipole field. This point occurred at $r = R/\alpha$, where α is the exponential decay rate of the field.

Consequently, the location was close to the sphere for a rapidly decaying field and far from the sphere for a slowly decaying field.

Another example was a current ramp that was curved at both ends in the shape of a symmetric cubic polynomial. In this case, the symmetry caused the O-type neutral point to gradually move to the center of the bubble, like the bubble of a sudden turn-on, but with leading and trailing sheaths of thickness $2R$. As in all of these problems, the O-type neutral point was a sink of flux for the on-bubbles and a source of reversed-polarity flux for the off bubbles.

17.6 Other Examples of Interacting Fields

Here, we looked for other inter-penetrating fields that might show the stretching and Pinchoff of loops. A trivial case was the sudden reduction of the strength of the current without turning it all the way off. This is equivalent to our previous case of turning the current off, but in the presence of a stronger background field. As in that previous case, the Pinchoff occurs in the leading sheath, but farther from the sphere and on a contour of less magnetic flux. Likewise, the bubble of reversed polarity forms farther from the sphere, before leaving a residual dipole field in its wake.

A non-trivial example consisted of a plane wave moving past a dipole field in two dimensions, somewhat analogous to the solar wind moving past the magnetic field of Earth. In this case, the familiar stretching and detachment that occurred on the outbound path also occurred on the inbound path, but in a reverse process. In particular, the incoming wave merged with the dipole field to form stretched loops, which then collapsed as the composite field became dominated by the dipole. A calculation showed that some of the field lines on the inbound side of the sphere were carried around to the outbound side—a result that does not make sense. How can an isolated dipole field end up with more flux on one side than the other when it lies in a vacuum?

This result was not caused by an error in the calculation. Instead, it was caused by the plane-wave model, whose electric field prevented A_z from being constant on magnetic field lines. To restore the dipole field to its original condition, two plane waves would have been required, one with a positive "meridional" field and the other with a negative field. And even this fix would leave the details of the instantaneous interaction with the dipole field subject to doubt. The spherical waves from distant spheres, like those in Chap. 13, provide a better description of the wave-dipole interaction.

The single, plane-wave model seems to be analogous to the infinitely long, straight current that I discussed in the Introduction. The current in a straight wire cannot produce the outgoing bubbles of circulating flux that one obtains from a current loop, either in a plane or on a sphere. Two, oppositely directed, linear currents would be necessary.

17.7 Short-Wave Radiation

In Chap. 5, the fields of a linearly increasing current ramp of finite length were derived by regarding the ramp as a long staircase of many small steps, and then adding the contributions of the individual steps. This procedure gave the desired fields, but these fields had different forms in different regions of space. There were oppositely directed meridional fields in sheaths of width, $2R$, on the leading and trailing boundary of a growing dipole field. The static field of a uniformly magnetized sphere was left behind this outgoing bubble of flux. For the linearly decreasing current ramp, the polarities of the sheath fields were reversed and the enclosed dipole field was collapsing rather than expanding. Except for the meridional field in the trailing sheath, no field remained behind this collapsing dipole.

These results were obtained for a ramp time greater than the transit time across the $2R$ diameter of the sphere (i.e., the "long-wave" solution). There is also a "short-wave" solution in which the ramp time is less than $2R$, but, I thought it would be simpler to pursue the implications of the long-wave solution first, and leave the "short-wave" solution until it was needed. This happened in Chap. 9, when the radiated flux and energy were derived as a function of the ramp time on the full interval $(0, \infty)$. The short-wave results were also needed in Chap. 10 where we compared the radiated energies for a variety of non-linear ramps.

For the long-wave field, the infinitesimal bubbles occurred in two, long chains that overlapped everywhere except in $2R$-wide regions at their ends. By superimposing these chains, we obtained oppositely directed meridional fields in the sheaths at the leading and trailing ends of the chains where the circulating contributions of the bubbles did not cancel out. Between these sheaths, the circulating fields cancelled out, leaving the static dipole fields that trailed behind them. Although these were static dipoles, their sum increased with time as more bubbles contributed to this central region. The result was a net dipole whose strength increased linearly with time. However, once the end of the chain had been reached, there were no more dipole contributions. The uncancelled circulating contributions produced the meridional field of the trailing sheath, and the accumulated static contributions of the bubbles left a static dipole field behind.

For the short-wave field, the infinitesimal bubbles (and their trailing dipoles) were shifted in time by an amount that was less than the $2R/c$ transit time that light would take to cross the sphere. In effect, the net contribution was an average of many closely spaced bubbles, and appeared like a single bubble with slightly blurred leading and trailing ends. Its field lines looked similar to those of the sudden turn-on, but with leading and trailing sheaths of thickness equal to the ramp time.

17.8 Radiated Energy

In Chap. 10, we calculated the radiated energy for a general current ramp using two different methods, and applied the results to a variety of ramps.

First, we computed the energy provided by the sphere when the electric field of the wave moves past the electric current on the sphere. Part of this energy ($(3/2)\mathcal{E}_0$), where $\mathcal{E}_0$ is the volume energy in the sphere) went into the volume energy of the magnetic field, and the other part escaped as radiation. This radiated energy was largest ($(3/2)\mathcal{E}_0$) for the sudden turn-on and decreased monotonically with the duration of the current ramp. It was $(6/5)(\Delta r)^{-2}\mathcal{E}_0$ for a linear ramp of duration, Δr, provided that $\Delta r > 2$.

However, for $\Delta r \ll 2$, the energy decreased linearly at a rate that was inversely related to the slope of the current profile. For steep profiles, the energies decreased slowly with ramp time and for gradual profiles, the energies decreased rapidly. Thus, the radiated energy decreased rapidly for the gradually decaying exponential profile, but more slowly for steeper profiles like the symmetric cubic and the linear profile.

Second, we computed the radiated energy by evaluating the Poynting flux across a given radius and then let this radius become infinite. To obtain the entire radiated energy, we needed to add the contributions from the leading and trailing sheaths (when they existed) as well as the central region between them. This allowed us to see where the dominant contributions originated. In general, the dominant contributions occurred where the current profile had the greatest "acceleration" or curvature. For the linear ramp, this was at the ends of the ramp, and for the quadratic/linear (Q/L) ramp, it was at the linear (L) end of the ramp.

For the symmetric cubic current profile, the dominant energy came from the central region, but even in this case the maximum occurred at the ends of the profile where the acceleration was the greatest. For more rapidly rising profiles, like the symmetric fifth-power and the symmetric seventh-power, the two peaks moved inward from the ends of the ramp to symmetrically placed positions closer to the middle of the ramp.

17.9 Oscillating Currents

In Chap. 11, we examined electric currents that are turned on and off repeatedly and found that the resulting bubbles may overlap, depending on the repetition frequency. This overlap caused the fields to interfere, and made the radiated power larger or smaller, depending on whether the interference was constructive or destructive.

For a linear ramp of duration, Δr, there was no interference provided that the repetition time, τ, exceeded $\Delta r + 2$ (in units of the time R/c). In this case, there was a space between the upward-directed meridional field in the trailing edge of the on-bubble and the upward-directed meridional field in the leading edge of the off-bubble. When this space was reduced to zero, these two upward-directed meridional fields lay side by side, providing a total energy of $2 \times (3/5)(\Delta r)^{-2}$. Because these two regions occurred at intervals of $\Delta r + 2$, the radiated power was $(6/5)(\Delta r)^{-2}/(\Delta r + 2)$, or approximately $(6/5)(\Delta r)^{-3}$ when $\Delta r \gg 2$.

However, if the current were turned off immediately at the end of the on-ramp, before the on-bubble had a chance to leave the sphere, then the leading sheath of the off-bubble would fully overlap the trailing sheath of the on-bubble. Consequently, the magnetic field would be twice as strong in this region, and its radiated energy would be four times as strong, and therefore $(12/5)(\Delta r)^{-2}$ in each non-overlapping sheath. Furthermore, the overlap caused this energy to be crowded into a slightly smaller duration, Δr compared to $\Delta r + 2$ for the bubbles that just touched, giving a radiated power of $(12/5)(\Delta r)^{-3}$. For large bubbles with $\Delta r \gg 2$, this power is approximately twice the power obtained from the bubbles that were just touching.

This enhanced power occurred because the energy was carried mainly in the sheaths and not in the middle region between them. For sinusoidal current profiles or otherwise smooth profiles like the symmetric cubic, whose energy was distributed within the middle region, the interference between the fields at the ends of the ramps had little effect on the radiated energy or power. So for those smoothly oscillating, low-frequency current profiles, the radiated power reduces to the classical value given by the square of the acceleration, which works out to $(1/2)A^2\omega^4$ for oscillating profiles of amplitude, A, and frequency, ω (Slater and Frank 1947).

However, the situation is quite different for short-wave radiation for which the repetition time is less than $2R/c$. In Chap. 11 we considered a square-wave current profile obtained by turning the current suddenly on and then suddenly off after a time, $\tau < 2$ (again in units of R/c). This process generated a periodic series of bubbles whose leading edges were separated by the amount τ and whose widths were 2 units of R/c.

The idea was to decrease τ and see how the radiated power changed as the overlapping meridional fields of the bubbles went in and out of phase. We assumed that the field had been oscillating forever, so that the chain of bubbles was infinitely long and there were no end effects. However, by decreasing τ, the density of these bubbles increased, and the number of overlapping bubbles increased according to $\nu = 2/\tau$. Thus, to study the interference, we increased ν through a series of integers and watched to see how the overlap changed. Figure 17.1 illustrates this effect using circular field lines to represent radiated bubbles of flux.

As ν started to exceed 1, the bubbles began to interfere constructively, with the trailing part of an on-bubble beginning to overlap the leading part of the next off-bubble. When $\nu = 2$, the distance between each bubble is $\tau = 1$, and the region of overlap (indicated by the black dot) contains upward-directed field lines that add in phase. If the number of circles were extended to form a long series, additional lens-shaped regions of overlap would appear in a corresponding series of strong fields of alternating magnetic polarity, like those shown in the upper right panel of Fig. 11.3. We found that the radiated power was 9/4 in units of the power $p_0 = \mathcal{E}_0 c/R$, which is three times the power radiated when the non-interfering bubbles are just touching.

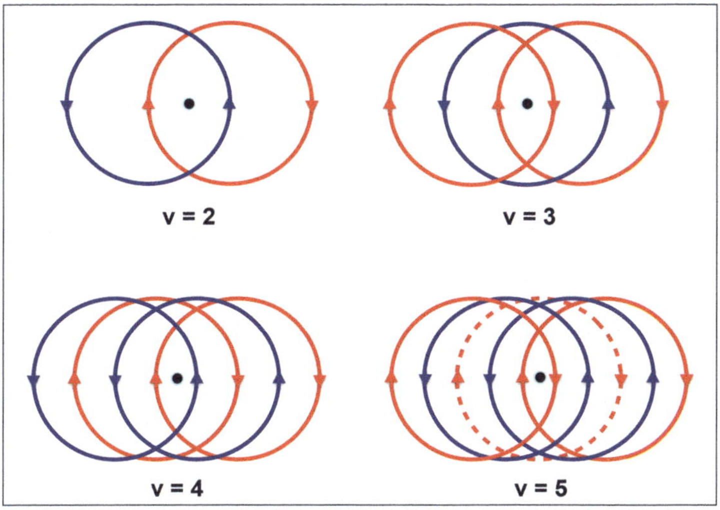

Fig. 17.1 Circular field lines of radius R, simulating overlapping bubbles of flux produced by an oscillating square wave of electric current on a sphere of radius R. In each case, the repetition time is τ and the number of overlapping circles is $\nu = 2/\tau$. A black dot marks the center of each lens-like region of overlap. If additional circles were added to each example, they would create additional lens-like regions of overlap in a series of alternating magnetic polarity with strong fields when ν is even and weak fields when ν is odd

As ν increased further, the overlap decreased and the radiated power began to fall. When ν reached 3, each field point received contributions from three bubbles—a blue off-bubble centered on the field point and two, red on-bubbles located symmetrically on each side. The central off-bubble probably provided little, if any, contribution, due to its symmetry about the field point, and the pair of on-bubbles probably had little effect because their meridional components at the field point were pointed in opposite directions. So the result would be a chain of very weak meridional fields, like that shown in the lower left panel of Fig. 11.3. Our detailed calculation showed that the radiated power reached a minimum value of $(3/4)\nu^{-2}$, which is only 1/12 when $\nu = 3$.

As ν increased further, the results repeated. At $\nu = 4$, the field point received contributions from four bubbles, with the first and fourth bubbles providing the main contribution because they passed closest to the field point and were oriented in the same direction. This led to a radiated power of 9/4, distributed in bands of alternating positive and negative polarity and with widths of $\tau = 1/2$, as shown in the lower right panel of Fig. 11.3. At $\nu = 5$, the field points received contributions from five bubbles. As shown in Fig. 17.1, the first and fifth bubbles passed closest to the field point, but were oriented in opposite directions. The second and fourth (blue) bubbles passed farther away and also were oriented in opposite directions. As for the case with $\nu = 3$, the field of the third bubble (dashed) was relatively weak at the field point, and so probably had little, if any, contribution. The net result was destructive interference with a radiated power of only $(3/4)(5^{-2}) = 0.03$.

The continuation of this process gives the periodic variation of power in Fig. 11.2, and shows that for a short-wave oscillating current source, the interference of the bubbles produces a drastic effect on the average radiated power. In particular, that power is a maximum whenever $2/\tau$ is an even integer and the consecutive bubbles obtain their maximum constructive interference. It was particularly interesting to interpret this result in terms of the wavelength of the oscillating current source, which is $\lambda = 2\tau R$ when the time between the turn-on and turn-off is $\tau R/c$. This means that $R = n(\lambda/2)$, where $n = 1, 2, 3, \ldots$, and is consistent with the well-known result that the radiated power from an antenna is maximum when the "length" of the antenna (radius of the sphere) is an integral number of half-waves. Now, we can understand this result as the interference between the trailing half of one bubble and the leading half of the next bubble when $n = 1$.

17.10 Dynamics

Up to this point, we have asked what the field would be like and how it would evolve with time without thinking about the effort that would be required to make it happen. In Chaps. 12 and 13, we examined the forces that would be involved and the resulting momentum of the outgoing bubbles. To do this analysis, we used the general principle that the stress on a closed surface is balanced by the force on the currents that are enclosed by that surface plus the rate of increase of momentum of the bubbles that are also enclosed by the surface.

We found that when the current on the sphere was turned on suddenly, there was a monotonic increase of force on the current while the bubble was being formed inside the sphere. This force was directed horizontally in all directions around the sphere (perpendicular to the polar axis, if we think of the sphere as a globe with north and south poles). This symmetry caused the net force on the sphere to be zero, but in any horizontal direction, it increased with time until the bubble left the sphere. At that time, the horizontal force dropped suddenly to $(21/16)\ F_0$ where $F_0 = (\pi R^2)(B_0^2/2\mu_0)$. This static value was also the average value that the force had while the bubble was being formed inside the sphere as r_L increased from 1 to 3.

Next, we surrounded the entire bubble, both inside and outside the sphere, with a closed surface and calculated the momentum as a function of time. The interior portion of the momentum varied strongly with time, first decreasing to a negative maximum as the trailing edge of the bubble moved inward and then changing sign and increasing to a positive maximum as the trailing edge turned around and headed outward toward the surface of the sphere again. By comparison, the exterior portion was always positive and its temporal variation was relatively weak. Consequently, the temporal variation of the entire bubble was dominated by the temporal variation of its interior portion, and showed a strong variation, first toward the center and then away from the center as time increased from $r_L = 1$ to $r_L = 3$. However, this tendency was not perfect and the momentum did not change sign exactly when the bubble turned around at $r_L = 2$.

Also, from an approximate solution, we found that the force on the electric current initially equalled the rate of decrease of the total momentum of the bubble with respect to time, so that the net stress on the bubble vanished. In other words, for a while, the bubble increased its inward momentum by pushing on the electric current without any help from the surrounding field. However, as time passed, the net stress on the bubble increased, and the rate of decrease of the momentum was no longer balanced by the force on the electric current.

After leaving the sphere at $r_L = 3$, the momentum slowly approached an asymptotic value of $(9/16)F_0 = \{27/64\}\mathcal{E}_0/c$ as $r_L \to \infty$. So the bubble carries away a momentum $(27/64)\mathcal{E}_0/c$ in each horizontal direction as well as a total energy $(3/2)\mathcal{E}_0$. This raised the question of whether other radiation fields might have the same value of energy per unit of momentum, $(9/32)c$, which might therefore be used to deduce their momenta from a knowledge of their radiated energies. This question was answered in the affirmative for the field of the linear current ramp whose radiated energy was $(6/5)(\Delta r)^{-2}\ \mathcal{E}_0$ and whose momentum was $(27/80)(\Delta r)^{-2}\ \mathcal{E}_0/c$. Like the radiated energy, all of the momentum was carried in the leading and trailing sheaths.

An interesting result occurred when we evaluated the stress on a thin sheath surrounding the leading edge of the bubble. We obtained a net value of $+(27/32)F_0$, which had to be the rate of increase of the momentum at the leading edge because no current was located in that sheath. Thus, despite the lack of momentum in such a thin sheath, there was an appreciable stress corresponding to a finite rate of change of momentum with respect to time. The same result was found inside the sphere at the leading edge of the ingoing wave. In Chap. 13, we recognized that such a bubble would exert a force when it collided with a superconducting wall.

17.11 Two Non-concentric Sources

In Chap. 13, we extended our analysis to two identical current sources of radius, R, and center-to-center separation, $2a$. We turned the currents on suddenly and simultaneously as seen from an observer located midway between them.

When the dipole moments were parallel, the outgoing bubbles passed through each other and left a composite field with flattened field lines at the mid-plane between them. We examined the stress on this plane in two ways. First, we regarded the mid-plane, $y = 0$, as a superconducting wall, and supposed that the fields in the $(y > 0)$-region evolved as if the sphere at $y = -a$ were not present. In this case, the stress on the mid-plane equalled the force on the induced current on the wall and our objective was to determine this force as a function of time. Second, we regarded the mid-plane as the face of a very large closed surface through a field-free region that enclosed the sphere at $y = +a$ and all of the fields around it. In this case, the stress equalled the increase of momentum in the region where $y > 0$ plus the force on the sphere at $y = a$ after the bubble from the other sphere had passed it.

The force on the wall had two parts—one due to the collision with the bubble and the other due to the composite dipole left behind. As the strong meridional fields in the leading and trailing parts of the bubble passed the mid-plane, the stress increased, decreased, and increased again in a cubic-like temporal profile. This final increase stopped when the trailing edge of the bubble passed through the wall, and the stress began its final decay toward zero, as the circular region of contact between the bubble and the wall moved progressively outward from its center. As the composite dipole field filled this region, it began to exert a force against the wall. This force increased toward an asymptotic value of $(3/8)a^{-4}\ F_0$ as the circular region of contact widened. However, this high inverse power of the half-separation, a, caused the force to be very small, unless the spheres were almost in contact and $a \approx 1$. Thus, for large values of a, the force of the composite dipole field was negligible compared to the force exerted by the bubble as it collided with the wall.

Next, we discarded the imaginary wall and regarded the mid-plane as the left face of a very large surface that enclosed all of the fields where $y > 0$. We changed the sign of the stress that we had previously computed with the wall in place, and equated that stress to the rate of increase of momentum in the $(y > 0)$-region. (The force on the sphere at $y = a$ vanished prior to the passage of the bubble from the left sphere, and, as mentioned above, even that force could be neglected after the passage of the bubble, unless the spheres were almost touching.)

Because the $(y > 0)$-region gained momentum from the rightward bubble at the same rate that it lost momentum from the leftward bubble, the stress at the mid-plane gave twice this value, which we called $2\dot{p}_c$. Because we had already obtained approximate expressions for the stress as a function of time, r_L, it was a simple matter to integrate this stress and obtain an approximate expression for p_c. This momentum increased rapidly with time, hardly showing the double exertion of stress that occurred when the bubble crossed the mid-plane, and then asymptotically approached the value $(9/16)p_0$.

When the dipole moments were anti-parallel, the bubbles left a composite field with an X-type neutral point in their common wake. This neutral point (or line in three dimensions) separated closed field lines of the separate spheres from field lines that now joined the spheres. As time passed, the amount of interconnected flux increased asymptotically toward a maximum value of $\Phi/\Phi_0 = 2/(\pi a)$, where Φ_0 is the flux of the uniform field within each sphere. Thus, as one would expect, the amount of flux linked between the two spheres decreased with the distance between the spheres.

For this anti-parallel configuration, the stress at the mid-plane was similar to that of the parallel configuration. There was another double exertion of stress when the alternating positive and negative meridional fields of the bubbles crossed the mid-plane, followed by the asymptotic increase of the stress from the composite dipole field. Whereas the parallel configuration gave a stress of $-(3/8)a^{-4}$, corresponding to a repulsive force between the dipoles, the anti-parallel configuration gave a stress of $+(3/8)a^{-4}$, corresponding to an attractive force. However, for widely separated dipoles with $a \gg 1$, both of these forces were negligible compared to stress produced by the bubbles, which corresponded to the same, outward-directed momenta for both configurations.

An interesting aspect of this two-sphere geometry was that the passage of each bubble past the dipole field of the other sphere was similar to the example of the plane wave moving past a two-dimensional dipole field that we considered in Chap. 8.2. More precisely, this two-sphere problem was similar to a hypothetical plane-wave problem in which the plane wave was followed by another plane wave of opposite magnetic polarity chosen to restore the dipole field to its original condition. The plots showed the characteristic stretching and Pinchoff that occurred on the outbound path prior to the separation of the bubble from the dipole field and the lack of such stretching on the inbound path. This difference between the loop configurations on the two paths was due to the differing gradients of the fields in the bubble and in the target dipole on these two paths. This did not occur for the field of the plane wave or for the sheaths of a bubble produced by a linear current ramp.

17.12 Concentric Sources and Total Energy

In Chap. 14, we considered two spheres with the same center, but with different radii, R_o and R_i, for the outer and inner sphere. These spheres had the same current profiles, $i(\theta)$, but the profiles were scaled by the radii so that the total current on each sphere was $3(B_o R_o/\mu_0)$ and $3(B_i R_i/\mu_0)$. We turned these currents on suddenly, but at different times. First, we turned on the current in the outer sphere and then waited until the transient was gone and the uniform field appeared in its wake. This process required an energy of $3\mathcal{E}_o$, of which half went into the field and half escaped with the bubble.

Next, we turned on the current in the inner sphere. When the dipole moments of the two spheres were parallel, field lines of the uniform field passed through the inner sphere from pole to pole, linking those two fields together. More energy was required to produce this composite field than to produce the separate dipole fields. The total energy required for turning on the currents in the two spheres was $\mathcal{E}_{tot} = 3\mathcal{E}_o + 3\mathcal{E}_i + M_i B_o$, where the extra term is the product of the dipole moment of the inner sphere and the field strength of the uniform field in the outer sphere.

Because the bubble from the inner sphere left this extra energy in the composite field, the bubble lacked the energy it needed to escape from the system with the expected radiation energy of $(3/2)\mathcal{E}_i$, where $\mathcal{E}_i$ is the energy of a uniform field of strength, B_i, in a sphere of radius, R_i. Consequently, the bubble acquired the balance when it passed the current in the outer sphere and left with its allotment of $(3/2)\mathcal{E}_i$. (Later, when we turned on the current in the inner sphere first, we achieved the same result, except that the bubble from the outer sphere obtained the balance as it passed the current in the inner sphere.)

When the dipole moments were anti-parallel, less energy was required to create the field, again by the amount $M_i B_o$, so this energy was given to the outer sphere as the bubble from the inner sphere passed through the outer sphere and left the system. Thus, in general, the extra energy could be represented by the dot product, $+\mathbf{M_i}\cdot\mathbf{B_o}$, which is positive for parallel dipole moments and negative for anti-parallel dipole moments of these concentric dipoles.

For the anti-parallel configuration, the dipole field of the inner sphere became separated from the uniform field of the outer sphere. This separation occurred at a spherical surface where the radial component of the composite field vanished, as if the boundary were made of superconducting material.

The radius of this spherical surface was $R_i(B_i/B_o)^{1/3}$, which is greater than R_i if $B_i > B_o$, and less than R_o if $B_o R_o^3 > B_i R_i^3$, which is equivalent to saying that the dipole moments satisfy the relation $M_o > M_i$. Under these conditions, the extra energy, $M_i B_o$, was less than the smaller of the two energies, $3\mathcal{E}_i$ and $3\mathcal{E}_o$, and therefore was relatively small for spheres of greatly differing radii.

Returning to the side-by-side current sources of Chap. 13, we found that less energy was required to create the flattened field configuration of the parallel dipoles than to create the linked configuration of anti-parallel dipoles. So again, the linked configuration required more energy than the isolated configuration, just as it had for the concentric spheres. The extra energy was given by $+\mathbf{M_0}\cdot\mathbf{B_1}$, where $\mathbf{M_0}$ is the dipole moment of sphere 0 and $\mathbf{B_1}$ is the field of sphere 1 at the location of sphere 0. This energy was negative for parallel dipoles because the field from one dipole bent over and pointed in the opposite direction at the location of the other dipole. Likewise, this energy was positive for the side-by-side anti-parallel dipoles.

It might seem puzzling that the extra energy required to create the composite field is $+\mathbf{M}\cdot\mathbf{B}$. We might have expected the energy to be $-\mathbf{M}\cdot\mathbf{B}$. However, the plus sign occurs because it refers to all of the energy required to create the composite field, and not just the "interaction energy" of the dipole placed in an external magnetic field, as Feynman explained in his Lectures on Physics (Feynman et al. 1964). The interaction energy, $-\mathbf{M}\cdot\mathbf{B}$, is $+(3/16)a^{-3}\ \mathcal{E}_0$ for the parallel configuration of side-by-side dipoles, which agrees with the energy that we obtained in Chap. 13 when we integrated the force of repulsion, $(3/4)a^{-4}\ F_0$, between the flattened dipoles. Nevertheless, when all of the sources are considered, it takes less energy to create those flattened fields than it does to create the separate dipoles.

As examples of concentric dipole fields that become separated, we considered anti-parallel dipoles for which the radius of separation occurred at R_i and R_o. When the separation occurred at R_i, the uniform fields of the two dipoles cancelled out, leaving no field inside the inner sphere. The composite field was essentially the field of the outer sphere with a hole in its center. In this case, the extra energy was $-3\mathcal{E}_i$, so that the total energy became $\mathcal{E}_{tot} = 3\mathcal{E}_o$. This means that there was no cost for the field of the inner sphere and the total energy requirement was the same as that for the outer sphere alone.

When the surface of separation occurred at R_o, the dipole moments of the two spheres cancelled out, leaving no field outside the outer sphere. This composite field was essentially the dipole field of the inner sphere confined to a spherical region inside the radius of the outer sphere. In this case, the extra energy was $-3\mathcal{E}_o$, so that the total energy became $\mathcal{E}_{tot} = 3\mathcal{E}_i$. In effect, there was no cost for the field of the outer sphere, and the energy requirement was the same as that for the inner sphere alone.

To summarize both cases, when we eliminated one of the fields, we also eliminated its cost, so that the energy required for the composite field was just the energy required for the other dipole alone. We shall return to this obvious result later when we summarize the calculations of the dipole field with a superconducting core.

17.13 The Boundary-Value Approach and the Conducting Core

At this point, we used a different method for calculating the fields. Instead of calculating the fields from a knowledge of the source currents, we began with spherical waves that satisfy the wave equation for a dipole field and looked for combinations of these waves that would reproduce the time variation of the radial field on the surface of the sphere, or equivalently, the time variation of the amount of magnetic flux that crosses that sphere. Such combinations would then give the fields at any radial distance.

We began with outgoing waves, $(1 - kr)e^{k(r-r_L)}$, and ingoing waves, $(1 + kr)e^{-k(r+r_L)}$, where in each case, k is the corresponding wave number. The idea was to find a distribution of wave numbers, $A^{out}(k)$ or $A^{in}(k)$, and to fold it together with the respective outgoing or ingoing waves to match the flux variation at $r = 1$. When the flux variations were relatively simple functions of time (i.e., polynomial functions of r_L), the distributions were combinations of Dirac delta functions and their derivatives, which were easy to combine with the spherical waves to obtain the fields everywhere.

We tested this approach for the sudden turn-on using the known time variation of $f(1, r_L)$ (which is the magnetic flux crossing the surface at $r = 1$), and easily recovered $f_o(r, r_L)$ and $f_i(r, r_L)$, which we had previously derived from the current source with some effort. However, we could not have done this without knowing the time variation of $f(1, r_L)$, which we had already learned by deducing the fields from the current. So the wave-based technique depends on knowing the transient variations of the radial component of the field on the surface of the sphere, which may not be known in general.

As a second example, I assumed that the surface variation was a simple monotonic function represented by a symmetric cubic polynomial for r_L in the interval (1,3). An interesting property of the resulting solution was that the current on the sphere had a profile of the form $1 - e^{-(r_L-1)}$, which rises from 0 to 1 with a time scale of 1 unit of R/c. Out of curiosity, I considered currents with different exponential rise times to see how their f values at $r = 1$ compared with the symmetric cubic profile and the peaked profile of the sudden turn-on of current.

The result was an interesting trend. For current rise times less than 1 (rates $\alpha > 1$), the f-values had profiles similar to that of the sudden current turn-on with peaks occurring before the fluxes returned to their final values. For very short rise times, the profiles approached the peaked distribution of the sudden turn-on. For rise times near unity, they approached the monotonic variation of the symmetric cubic profile. In contrast, for rise times greater than 1 (rates $\alpha < 1$), the f-values pulled away from the symmetric cubic profile and approached their asymptotic value of 1 at a slower rate. Perhaps the most revealing aspect of this comparison was that current profiles with short rise times produce temporary excesses of flux that must be removed from the sphere prior to attaining the final allotment of $\Phi_0 = (4\pi R^3/3)(B_0^2/2\mu_0)$. By comparison, currents with long rise times produce their flux monotonically and do not have the problem of removing excess flux from the sphere.

Our main use of the wave-based approach was to solve the problem of the conducting core in Chap. 16. In that problem, we placed a small superconducting sphere of radius ϵ at the center of a larger sphere of radius 1 (in units of R), and then turned the current on suddenly in the larger sphere. In order to satisfy the boundary condition that $B_r = 0$ on the inner sphere, we had to add an extra term with $k = 1$ to our "continuum" distribution, $A(k)$, for the second, outgoing wave. The continuum part of this spectrum satisfied the boundary condition at $r = \epsilon$, and the "line" at $k = 1$ satisfied the requirement that the outgoing wave have a sharp leading edge. This extra term corresponded to an exponential decay, so that after the passage of the second outgoing wave, the f-values were not static, but decayed toward their asymptotic values very quickly on a time scale of $\epsilon R/c$.

The asymptotic f-values were $1 - \epsilon^3$ outside $r = 1$ and $r^3 - \epsilon^3$ between $r = \epsilon$ (where the fields vanished) and $r = 1$. These f-values correspond to the fields of anti-parallel, concentric dipoles of equal strength and radii of $r = \epsilon$ and $r = 1$ when a conducting core is not present. So one effect of the conducting core is to introduce a rapidly decaying transient term to the anti-parallel fields of the concentric spheres.

Two other effects are to produce an X-type neutral point behind the second outgoing wave front, that marks the trailing edge of the outgoing bubble, and to remove the O-type neutral point for values of $\epsilon > 1/2$. As time increases, the X-type neutral point drifts farther behind the second wave front (causing the bubble to increase in size), while still moving outward from the sphere. As ϵ increases from 0 to 1/2, the asymptotic location of the O-type neutral point moves from the midpoint between the two outgoing waves to the location of the second wave. At $\epsilon = 1/2$, this O-type neutral point changes to a discontinuity where the meridional field component of the bubble suddenly reverses its direction.

There is a similar dependence of the radiated flux on the radius of the conducting core. If $\epsilon \leq 1/2$, the radiated flux is always $(3/4)\Phi_0$, but if $\epsilon \geq 1/2$, then the radiated flux is progressively less and becomes 0 when $\epsilon = 1$. In contrast, the flux inside the sphere asymptotically approaches $1 - \epsilon^3$ for all values of ϵ. Consequently, if the radius of the conducting core is small, then the flux left in the sphere (and in the external dipole) is essentially the same as when the core is not present. Likewise the radiated flux is the same as when the core is not present, unless the radius of the core is greater than 1/2. For large cores, the radiated flux is relatively small.

The energy budget does not share this strange dependence on whether ϵ is less than 1/2 or greater than 1/2. The total energy required to produce the field is $\mathcal{E}_{tot} = 3(1-\epsilon^3)\mathcal{E}_0 = 3\mathcal{E}_{out} - 3\mathcal{E}_{in}$, where these latter two terms are the energies required to create the separate fields of the outer sphere and the inner sphere (if its current were scaled to its radius and if it lacked the conducting core). Note that this total energy is different from the total energy required to create the composite field of the anti-parallel concentric dipoles of equal strength, which is $\mathcal{E}_{tot} = 3\mathcal{E}_{out} + 3\mathcal{E}_{in} - 3\mathcal{E}_{in} = 3\mathcal{E}_{out}$. More energy is required to create the anti-parallel dipoles than to create the dipole field with the conducting core.

The radiated energy for this sudden turn-on is $\mathcal{E}_{rad} = (3/2)(1 - \epsilon^3)\ \mathcal{E}_0 = (1/2)\ \mathcal{E}_{tot}$. Thus, half the energy is stored in the field and half is carried away in the outgoing bubble, just as if the conducting core were not present. Moreover, the volume energy is distributed with approximately $1 - (1/2)\epsilon^3$ inside the sphere and approximately $(1/2) - \epsilon^3$ outside the sphere, in each case, accurate to terms of order ϵ^6. Thus, for small ϵ, the volume energy distribution is similar to the distribution we obtained for the single source, with 1 unit of energy in the uniform field inside the sphere and 1/2 a unit in the dipole field outside the sphere. But when $\epsilon \to 1$ both the total energy and the radiated energy approach zero.

Chapter 18
Epilogue

My original plan was to find a current source that I could use to understand how magnetic fields change with time. The idea was to keep it simple, leaving out complications due to the presence of conducting material, and to derive the fields in closed form directly from Maxwell's equations. If I could do that, then I would have something trustworthy that I could refer to whenever I encountered problems about changing magnetic fields, as in seminars and published papers. It would be another reality check that I could use in the world of scientific research.

The spherical current source with the $\sin\theta$ distribution (θ is co-latitude in a spherical coordinate system) turned out to be ideal. The model was essentially the same as a flat current loop, except that it gave simple analytical solutions for the fields and their properties. In fact, its dipole field simulates the fields of many current sources at large distances, and therefore has the potential for contributing to a general understanding of changing magnetic fields.

What did I learn from this experience? I learned a lot about magnetic fields and electromagnetic waves, ranging from the details about magnetic fluxes, energies, momentum, and forces to interesting concepts that arose about the nature of magnetic field lines, the difference between total energy and interaction energy, and the bizarre analogy between the motions of magnetic field lines and the scattering of electrons as seen in Feynman diagrams (Feynman 1949, 1985).

Basic results included the way that magnetic fields emerge as the current turns on, with ingoing and outgoing waves creating a bubble of circulating magnetic flux, and the way that these fields disappear when the current is turned off. After solving the problem for currents that turned on or off suddenly, I extended the solution to include currents that change gradually, beginning with a simple linear increase and then going to more general changes including exponential variations and oscillating currents. The duration of the current ramp compared to the light transit time across the sphere turned out to be important because it marked the dividing line between long-wave radiation and short-wave radiation. The short-wave radiation occurred

N. R. Sheeley, Jr., *Transient Magnetic Fields*,
https://doi.org/10.1007/978-3-030-40264-8_18

when the duration was less than the transit time and led to interference effects when the current source was repetitive. This had a major effect on the radiated power.

However, the central theme was the two-wave character of the radiation, which ensured the formation of a neutral point where the legs of stretched loops could pinch together and detach the outgoing bubble from the dipole field that was left rooted in the sphere. This process would not have occurred for an idealized line current, and shows the importance of choosing the finite spherical current loop as a model.

The idealized line current is analogous to the plane wave that I used to study the reconnection that occurs when one magnetic field moves past another magnetic field. In this case, the plane wave moved past a two-dimensional dipole field, transferring flux from the inbound side of the sphere to the outbound side. Although this is similar to what happens when the solar wind sweeps past the Earth's magnetic field, it is contrary to what would happen in a vacuum. In a vacuum, the dipole field would return to its initial state after the passage of the wave.

Although this inconsistency surprised me, I should have expected it because the electric field ($E_z = -\partial A_z/\partial r_L$) in Eq. (8.25) prevented the contours of A_z from being constant on magnetic field lines. The addition of a second plane wave, following the first wave with oppositely directed magnetic field, might have corrected this flaw by restoring the dipole to its original condition. However, there would still have been the question of what happens during the interval between the two plane waves as well as during the encounter with the dipole field.

A better solution was to use the example in Chap. 13 where two identical current sources were placed side-by-side. In that case, the bubble from sphere 1 passed through the bubble from sphere 2 and then encountered the dipole field of sphere 2. A similar encounter occurred when the bubble from sphere 2 moved past the dipole field of sphere 1. In each case, the interaction was reversed on the inbound and outbound sides of the dipole: on the inbound sides, erosion of the dipole field lines preceded the stretching and collapse of loops, and on the outbound sides, collapse preceded the stretching and reconnection of field lines. However, when the current was turned on suddenly, the field of the bubble had a gradient along the direction of motion. This gradient combined with the field of the target dipole to give different loop configurations on the inbound and outbound sides of the dipole. When the dipole moments were parallel, the loops were elongated on the outbound side, but not on the inbound side. When the moments were anti-parallel, the fields were reversed so that the elongation occurred on the inbound side, but not on the outbound side. This inbound/outbound asymmetry did not occur during the passage of the plane wave in Chap. 8 because the field of that wave had a front/back symmetry.

Another surprise occurred when I made space-time maps of the magnetic field lines with the slit oriented along the "equatorial" radius to create tracks of the rising and falling loops. Low-lying loops with large flux contours expanded upward and then collapsed downward again, and their space-time tracks showed a corresponding rise and fall. For progressively greater heights and smaller contours, the collapse became faster until a space was produced where the legs of the remaining loops pinched together and detached the outgoing bubble from the dipole field. During this time, the return portion of the space-time track steepened and eventually turned backwards to form a time-reversed segment before turning forward again and continuing on. Thus low loops with large flux contours collapsed back toward the sphere, but loops, that had expanded too far, returned their fluxes by means of a reconnection process that is remarkably similar to the pair production that occurs in electron scattering at energies greater than the total rest mass of the electron-positron pair (Feynman 1949, 1985).

A third surprise occurred when I calculated the energy requirements for concentric and side-by-side spherical current sources. In each case, the total energy required to create the fields was $\mathcal{E}_{tot} = (3/2)\mathcal{E}_1 + (3/2)\mathcal{E}_2 + \mathcal{E}_{rad} + \mathbf{M_i}\cdot\mathbf{B_o}$, where $\mathcal{E}_1$ and $\mathcal{E}_2$ are the field energies of the uniform fields inside the two spheres, and $\mathcal{E}_{rad}$ is the total radiated energy from the two current sources (which depends on their respective time profiles). Thus, $(3/2)\mathcal{E}_1 + (3/2)\mathcal{E}_2 + \mathbf{M_i}\cdot\mathbf{B_o}$ is the volume energy in the composite field. Also, $\mathbf{M_i}$ is the dipole moment of the inner sphere (if the dipoles are concentric), and $\mathbf{B_o}$ is the magnetic field of the outer sphere at the location of the inner sphere. (If the spheres are side-by-side, the extra term becomes $\mathbf{M_1}\cdot\mathbf{B_2}$ or $\mathbf{M_2}\cdot\mathbf{B_1}$ for spheres at locations 1 and 2.) In each case, the total energy was largest when the dipole fields were linked together, and it was smallest when the dipole fields avoided each other (as if sheets of conducting material were present).

However, if we were to regard one dipole as a "test magnet" placed in the external field created by the other dipole, then the "potential" energy of that test magnet (its so-called interaction energy) would have been $-\mathbf{M_i}\cdot\mathbf{B_o}$. In this case, the flattened composite fields of the parallel, side-by-side dipoles would push those dipoles apart. Conversely, the linked composite fields of the anti-parallel dipoles would pull those dipoles together. This result illustrates the difference between the total energy required to produce the two-source configuration and the interaction energy of one dipole placed in the field of the other.

Figure 18.1 illustrates the isolated and linked fields of a compass (or small dipole) placed in the Earth's quasi-dipole magnetic field. When the compass needle is forced to point south (top panel), its field becomes isolated from Earth's field, as if the compass were surrounded by a spherical sheet of superconducting material (indicated by the dotted circle in Fig. 18.1). However, if the needle is released and swings around to become aligned with the Earth's field, then the Earth's field lines squeeze together and pass right through the compass to form the linked configuration shown in the bottom panel. Consequently, the interaction energy (or dipole potential energy) changes from $E_{int} = +\mu B$ to $E_{int} = -\mu B$, where μ is the dipole moment of the compass magnetic field and B is the magnitude of the Earth's field at the location of the compass.

This final figure leads back to the Introduction where I began with simple magnets and a compass placed in Earth's magnetic field, and wondered about a simple model that might lead to a better understanding. The results described here provide the reference that I was looking for when I began to calculate the radiation field for a sudden increase of current on the surface of a sphere. The model is simple and easily applied to other current sources, particularly at large distances where most fields approach the field of a dipole. The vacuum model is applicable to radiation

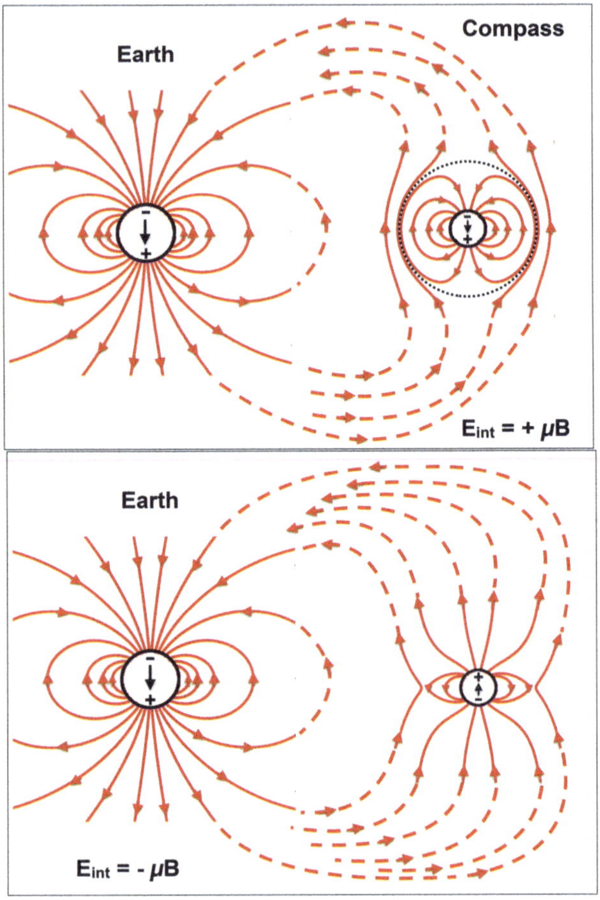

Fig. 18.1 Sketches, showing the field of a compass (or small magnetic dipole) placed in the Earth's "dipole" field. When the compass is oriented with its dipole moment, $\boldsymbol{\mu}$, pointing opposite Earth's field, **B** (top panel), the field of the compass becomes isolated, as if it were contained within a spherical, conducting sheet (dotted circle). When the compass needle swings around and points north (lower panel), its field lines become linked with the Earth's field, and its interaction energy, E_{int}, decreases

from an antenna. In particular, from our analysis of short-wave radiation from a repeating current source, we found that the radiated power was maximum when the radius of the sphere was one-half the wavelength of the radiation, analogous to the "half-wave" condition for the maximum power radiated from a radio antenna.

I originally thought that this research would be a first step toward understanding changes of the Sun's magnetic fields. Of course, the vacuum-field analysis cannot produce the magnetosonic waves that occur in a conducting fluid. However, the vacuum fields are easy to calculate and they do have waves when the displacement current is retained in Maxwell's equations. So, despite the much higher speeds of the electromagnetic waves (which are 10^2–10^3 times faster than the magnetosonic speeds), I wondered if these electromagnetic waves would provide clues for understanding the transients and magnetic field reconnections that remove stress from the Sun's corona.

There are cases in which vacuum fields have been used to understand fields in a conducting fluid. Parker (1994) has used the two-sphere "flattened-field" model in Chap. 13 to calculate the free energy of magnetic discontinuities formed in a conducting fluid. He noted that the method can be extended to a number of multipole fields provided that the multipoles are isolated from each other in separate "magnetospheres" with no vacuum field lines linking them together. Thus, the linked and isolated configurations shown above and discussed in Chaps. 13 and 14 are useful concepts for calculating the free energy of magnetic discontinuities in a conducting fluid, even though those configurations refer to vacuum fields!

Another case is the potential field model often used in solar physics to derive coronal magnetic fields from their observed or simulated sources of flux in the photosphere (Altschuler and Newkirk 1969; Hoeksema 1984; Schatten et al. 1969; Wang and Sheeley Jr. 1992). This model is essentially a vacuum-field model with the displacement current (and its associated waves) left out. I wondered if anything interesting could be learned about the field changes by putting the displacement current back into the model. If ongoing transients and magnetic field reconnections are responsible for keeping the coronal field current-free on the average, then perhaps the electromagnetic waves could provide some clues to the nature of this transient behavior, even if these waves were $10^2 - 10^3$ times faster than the relevant magnetosonic waves.

Regardless of whether this analysis provides any help in understanding the Sun's magnetic field, it has already improved my understanding of transient magnetic fields in a vacuum, revealing the two-wave character of the radiation and the formation and propagation of a magnetic bubble. In addition, the analysis has produced several interesting and unanticipated results. I hope that others will enjoy reading about them as much as I have enjoyed deducing them from Maxwell's equations. It is amazing how much is contained in those four equations.

References

Altschuler MD, Newkirk G (1969) Sol Phys 9:131
Feynman RP (1949) Phys Rev 76:749
Feynman RP (1985) QED: The strange theory of light and matter. Princeton University Press, Princeton, p 99
Feynman RP, Leighton RB, Sands M (1964) Feynman lectures on physics. Vol. 2: mainly electromagnetism and matter. Addison-Wesley, Reading, MA, p 15.5
Hoeksema JT (1984) Ph.D. Thesis, Stanford University
Jackson JD (1975) Classical electrodynamics. Wiley, New York, p 740
Leighton RB (1959) Principles of modern physics. McGraw-Hill, New York, p 409
Panofsky WKH, Phillips M (1962) Classical electricity and magnetism. Addison-Wesley, Reading, MA, p 245
Parker EN (1994) Spontaneous current sheets in magnetic fields with applications to stellar X-rays. Oxford University Press, New York, p 131
Schatten KH, Wilcox JM, Ness NF (1969) Sol Phys 6:442
Slater JC, Frank NH (1947) Electromagnetism. McGraw-Hill, New York, p 159
Wang YM, Sheeley Jr NR (1992) Astrophys J 392:310

N. R. Sheeley, Jr., *Transient Magnetic Fields*,
https://doi.org/10.1007/978-3-030-40264-8

Index

N. R. Sheeley, Jr., *Transient Magnetic Fields*,
https://doi.org/10.1007/978-3-030-40264-8

The manufacturer's authorised representative in the EU is Springer Nature Customer Service Centre GmbH, Europaplatz 3, 69115 Heidelberg, Germany. If you have any concerns regarding our products, please contact ProductSafety@springernature.com

Printed and bound by CPI Group (UK) Ltd, Croydon, CR0 4YY
30/09/2026
02225647-0002